GEOMETRIC SCULPTURES

An exploration of building blocks construction techniques

ROCKY LAU

To order additional copies of this book, contact
Toll Free +65 3165 7531 (Singapore)
Toll Free +60 3 3099 4412 (Malaysia)
www.partridgepublishing.com/singapore
orders.singapore@partridgepublishing.com

ISBN
ISBN: 978-1-5437-6483-3 (sc)
ISBN: 978-1-5437-6487-1 (hc)
ISBN: 978-1-5437-6482-6 (e)

Library of Congress Control Number: 2021909079

Print information available on the last page.

05/27/2021

PARTRIDGE

Author

Rocky Lau

Photography

Rocky Lau

Book Design

orcades:gn

PARTRIDGE

ABOUT AUTHOR

Rocky Lau is a mechanical design engineer, entrepreneur and inventor based in Singapore. He is a drummer with a local band playing classic rock and oldies. His interests are photography, collecting antique calculators and of course, playing and creating geometric sculptures.

This book is about the construction of geometric sculptures using mainly building blocks and toy construction parts. Most of these geometric sculptures are already well known and have been mainly constructed in wood, metal or plastic. They are also popular as origami models. Only a few of these are my own discoveries.

I have chosen this as a media because it is easy to assemble without the preparation of materials as compared to using wood. Building blocks is excellent for prototyping for these projects. Your ideas can be realized with minimum effort and easy to make changes and trying out other designs.

I hope you will have as much fun as I did in creating the sculptures and to solve problems during the assembly. At times it is similar to solving a puzzle and it needs patience.

A good compilation is from a book "Orderly Tangles" by Alan Holden. He use the term "regular polylinks " to describe these type of geometric linkages derived from polyhedra.

As regular polylinks become more widely known perhaps someone will discover significant applications for these beautiful mathematical objects. By using this as a media to create sculptures and models will hopefully make these more easily accessible to all.

Explore and have fun!

GETTING STARTED

To start making these sculptures you will need to buy a quantity of construction parts from online sources such as Bricklink. If you like to make the more complex models it will be helpful if you have access to a CAD program. There are online CAD viewer programs to see the model and to manipulate the views. There are a number of free online viewing programs or CAD software downloads. I think this will make the assembly process easier as the complex models are quite difficult to assemble without help from a CAD viewer. I have uploaded a number of my CAD models to GrabCad which you can download from the links provided. GrabCad do have a 3D viewer for the step files.

All uploaded CAD models at - https://grabcad.com/rocky.lau-1/models

Technic parts can be ordered from www.bricklink.com.

POLYHEDRA

Only five regular polyhedra exist: the tetrahedron (4 triangular faces), the cube (6 square faces), the octahedron (8 triangular faces), the dodecahedron (12 pentagonal faces), and the icosahedron (20 triangular faces).

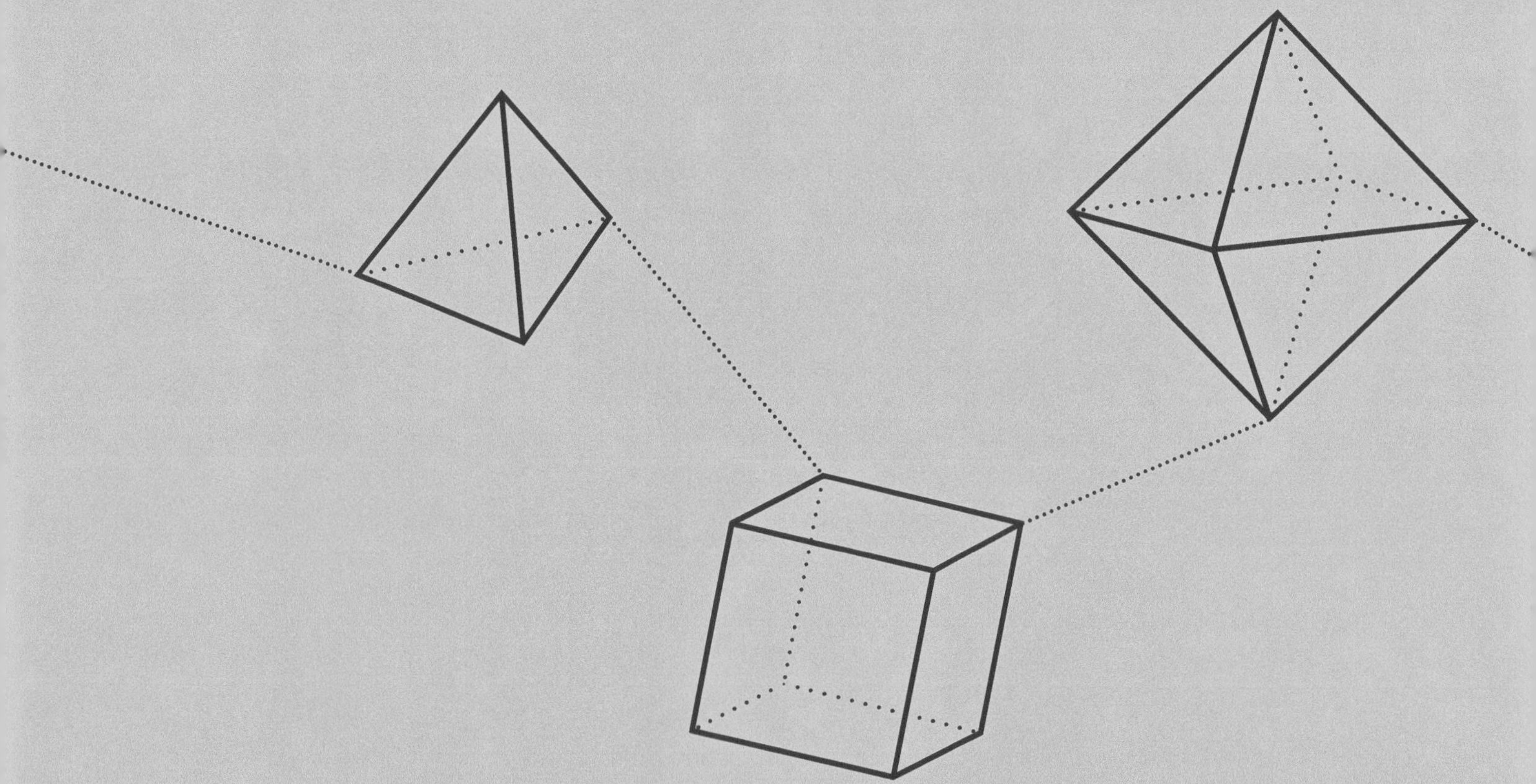

Tetrahedron Cube Octahedron

Any of these five geometric solids whose faces are all identical, regular polygons meeting at the same three-dimensional angles is a Platonic solid. The sculptures in this section are based on the Platonic solids and their derivations.

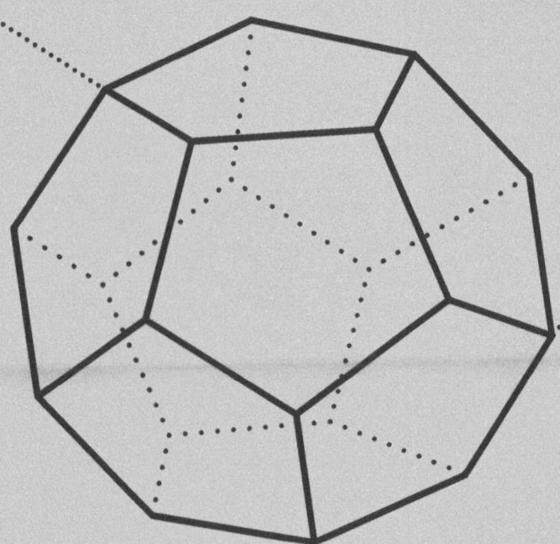

Dodecahedron

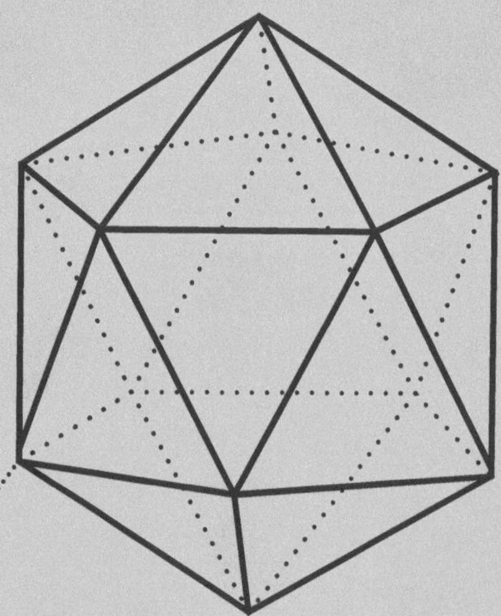

Icosahedron

CONTENTS

TETRAHEDRON

AND VISUALLY SIMILAR STRUCTURES

The Tetrahedron has 4 triangular faces, 6 straight edges, and 4 vertex corners. This is a construction method and the adjacent triangles forming double edges and vertices.

So this is not a true tetrahedron as represented by the construction method. This may be called as 4 triangles in a tetrahedron configuration. All the following sculptures will be using similar construction methods or its variations based on the polyhedra configuration.

CHAPTER 1

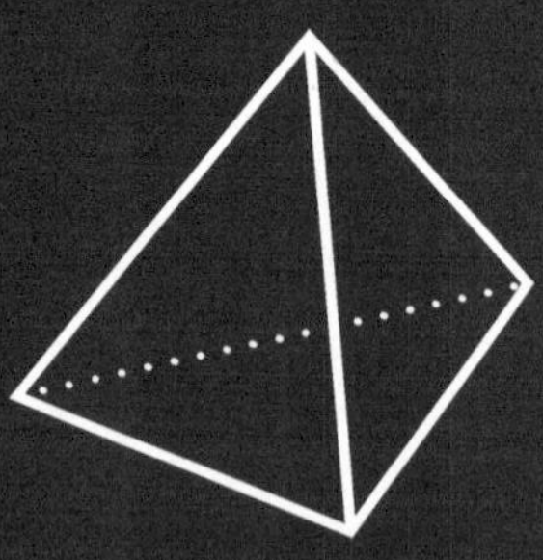

A tetrahedron formed by 4 triangles connected to each other with hinges.

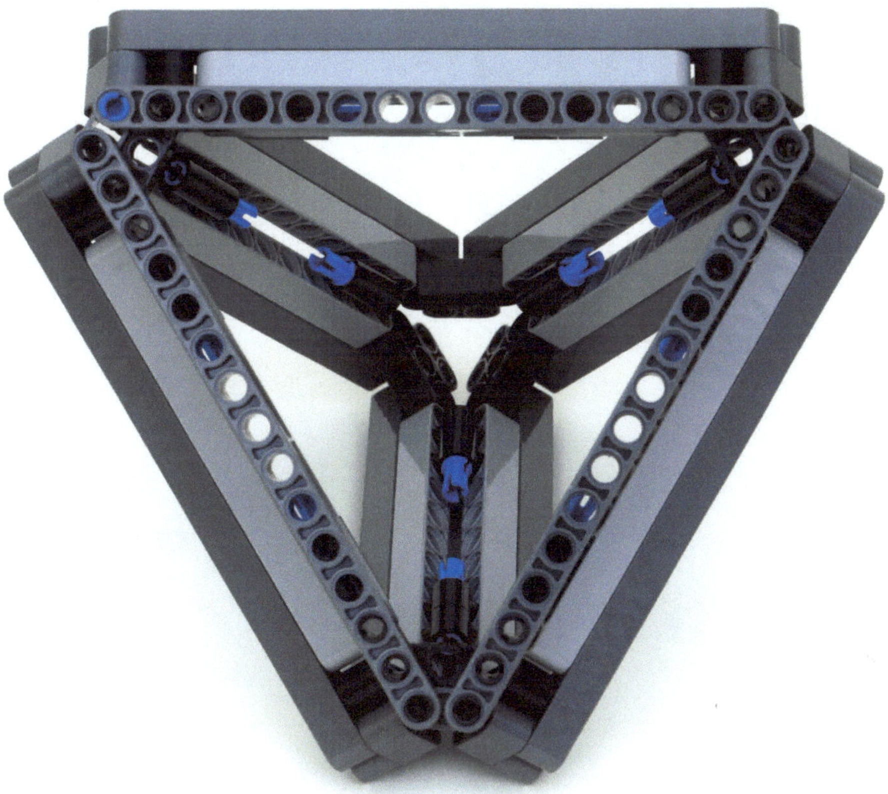

Parts:

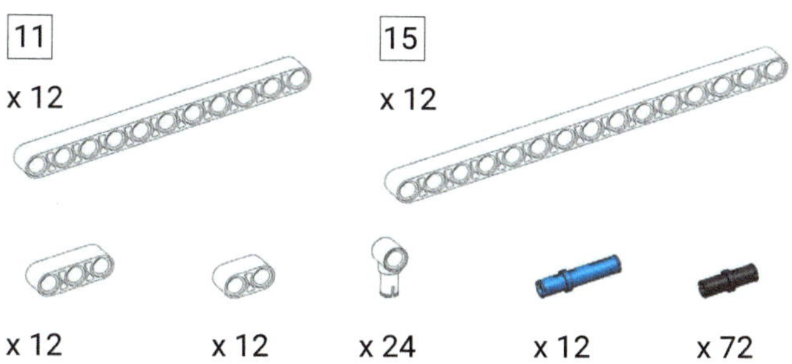

Hinge construction

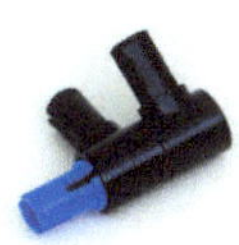

1 triangular face of the tetrahedron

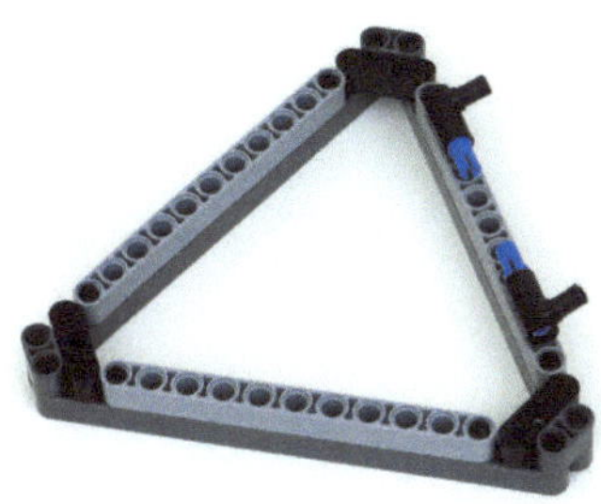

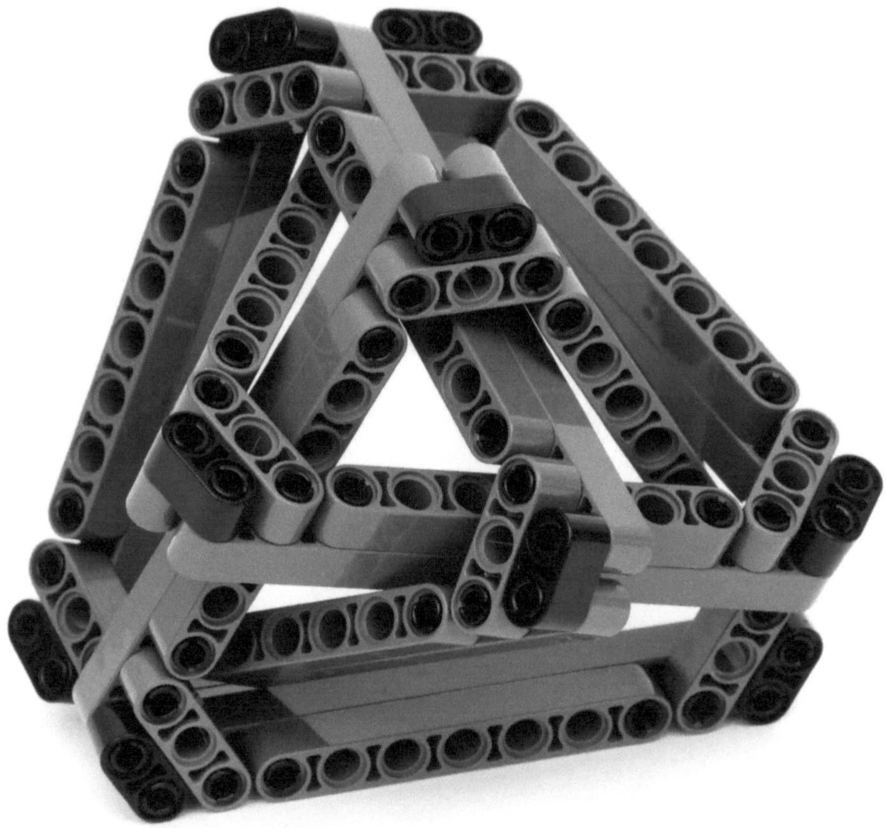

Three interlocking triangles in a Borromean link configuration - this is closely related to the Valknut. This is a good example of a Brunnian link - the three triangles are interlinked in such a manner that no two of them are linked but all three are simultaneously linked. If one is removed the other two comes apart.

Parts:

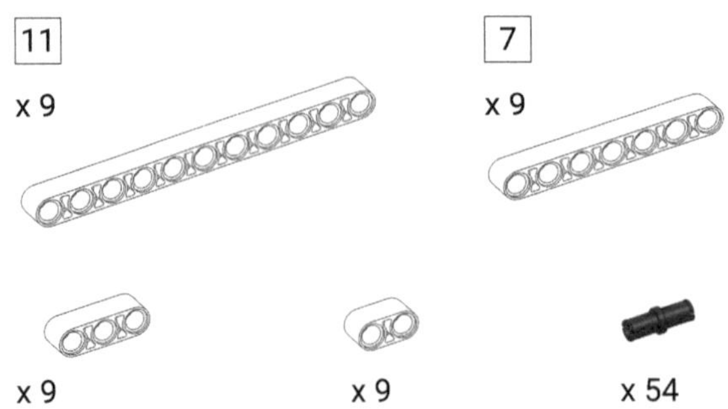

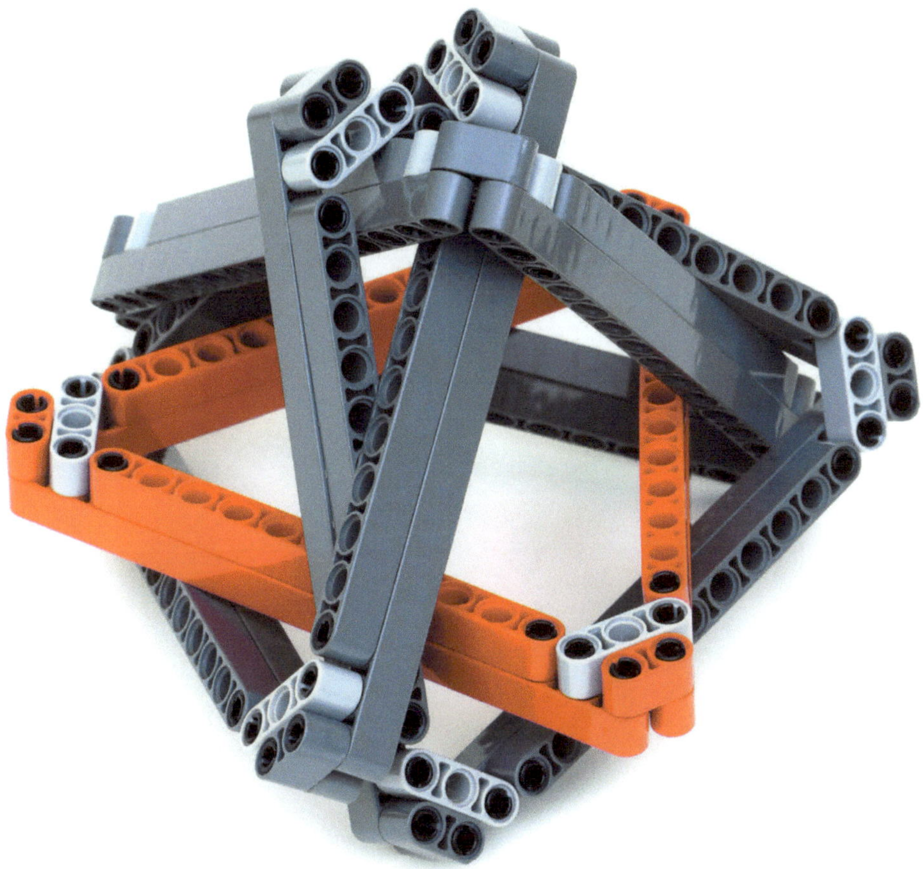

4 interlocking triangles - 3 triangles in a Borromean link configuration with a fourth identical triangle to hold it in place.The fourth triangle is a 'no link'.

Parts:

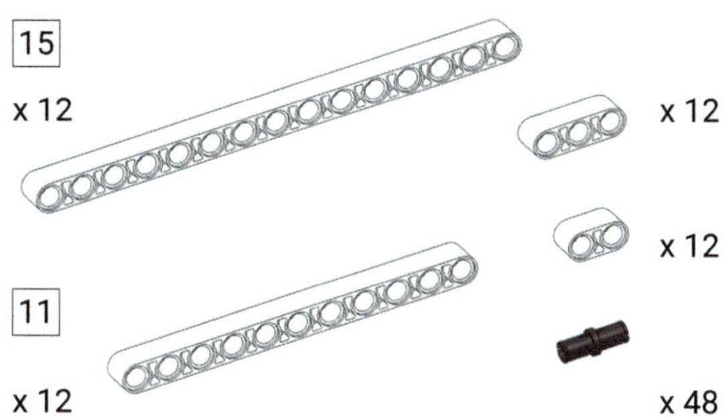

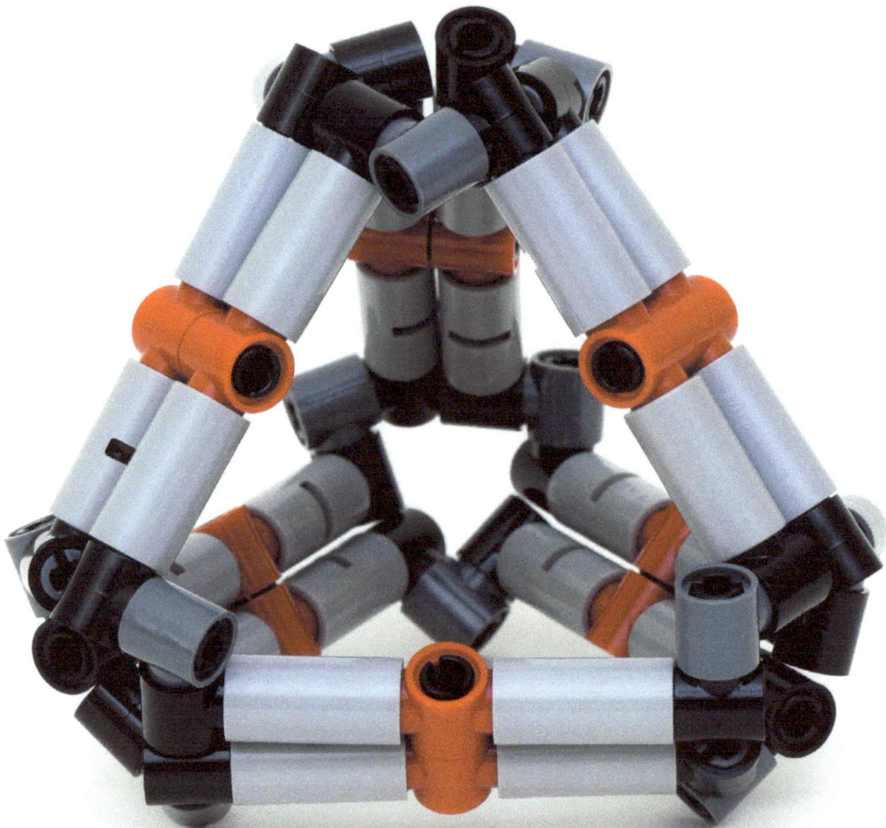

A tetrahedron configuration using connectors.

Parts:

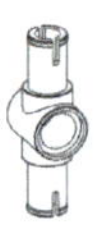

x 12

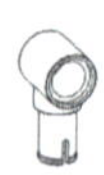

x 36

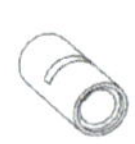

x 24

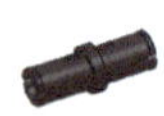

x 18

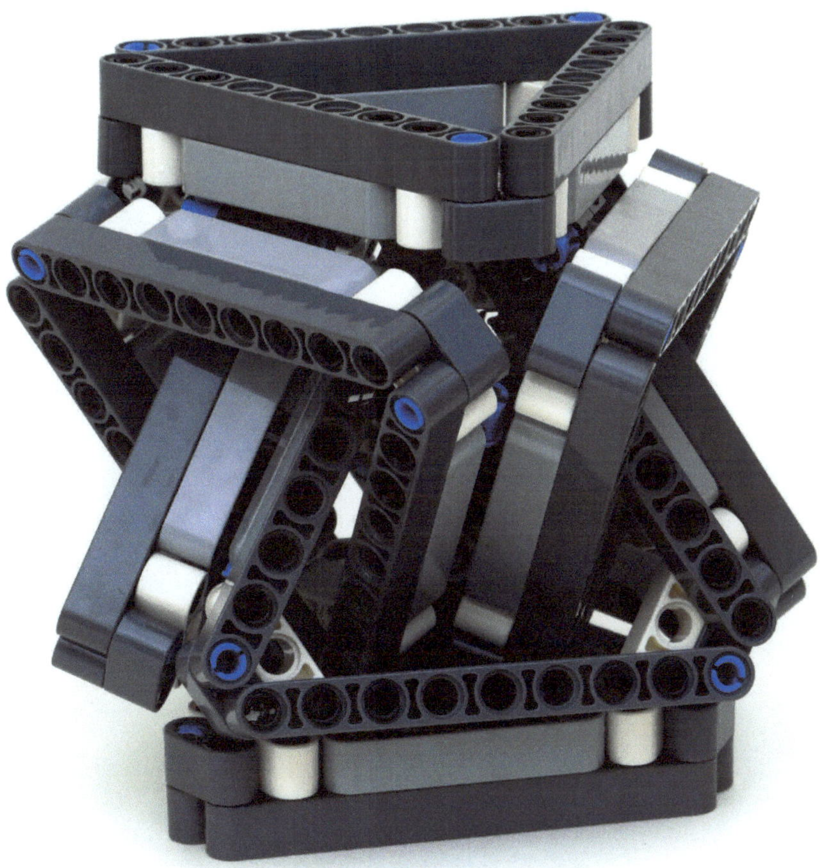

2 intersecting tetrahedrons - each tetrahedron is formed by 4 triangles.

1 Triangular face of the tetrahedron with a typical hinge shown.

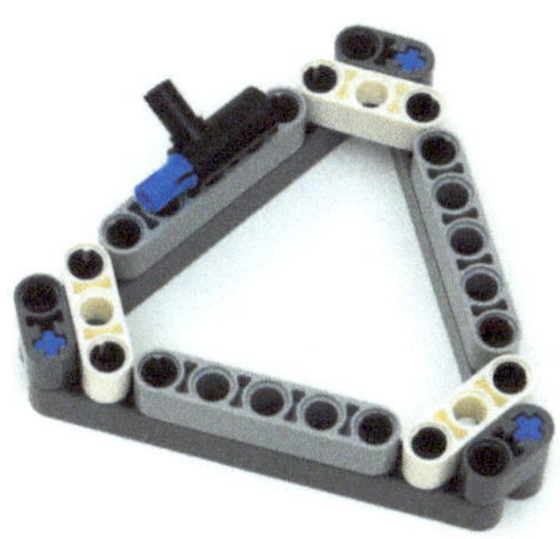

Parts:

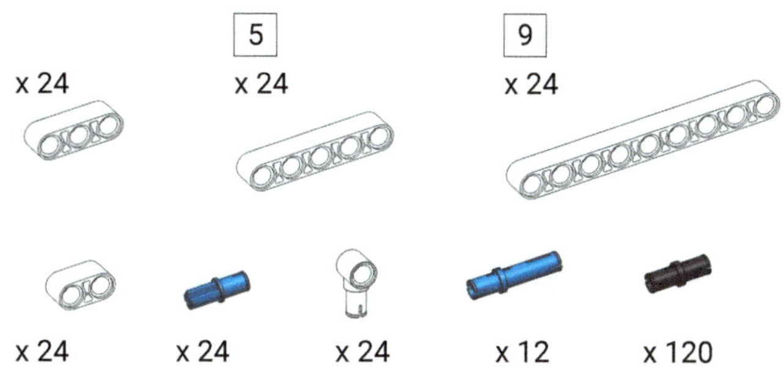

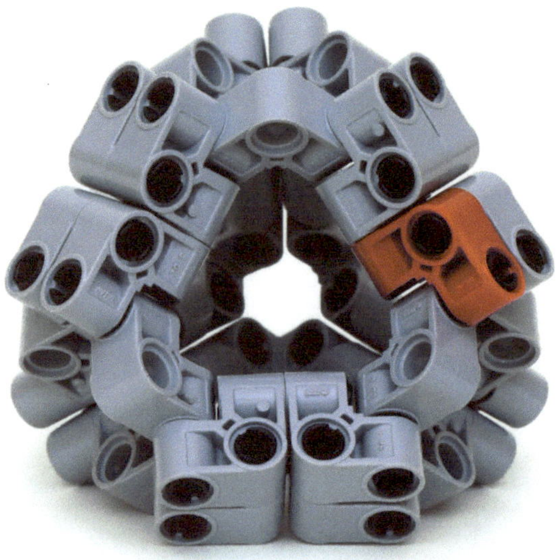

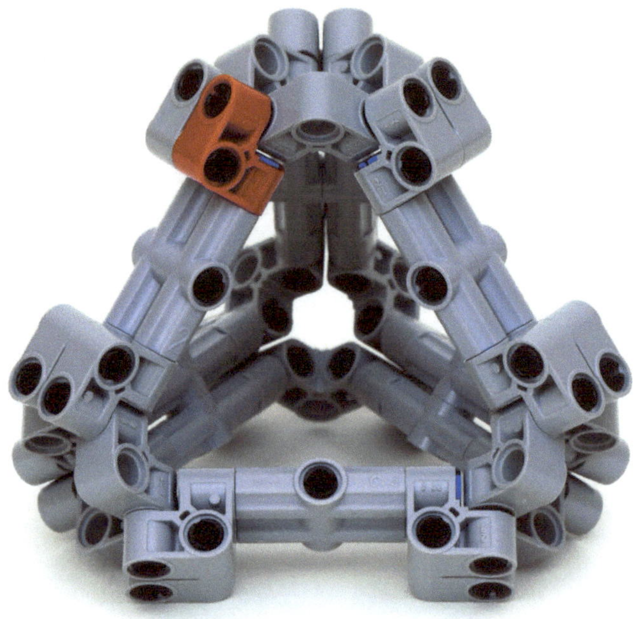

The left model is similar to the right model except having the center pairs of connectors removed.

A typical 'Vertex' construction.

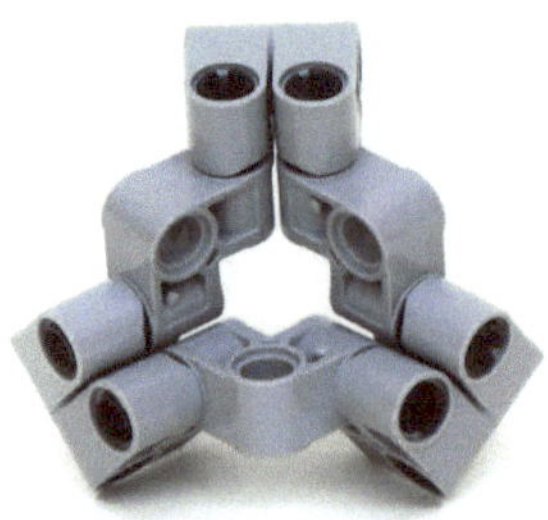

Parts:

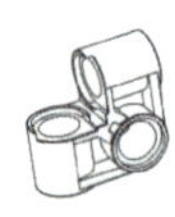

x 36

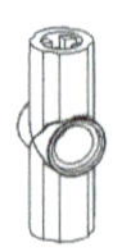

x 12

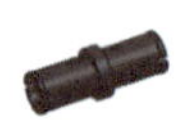

x 66

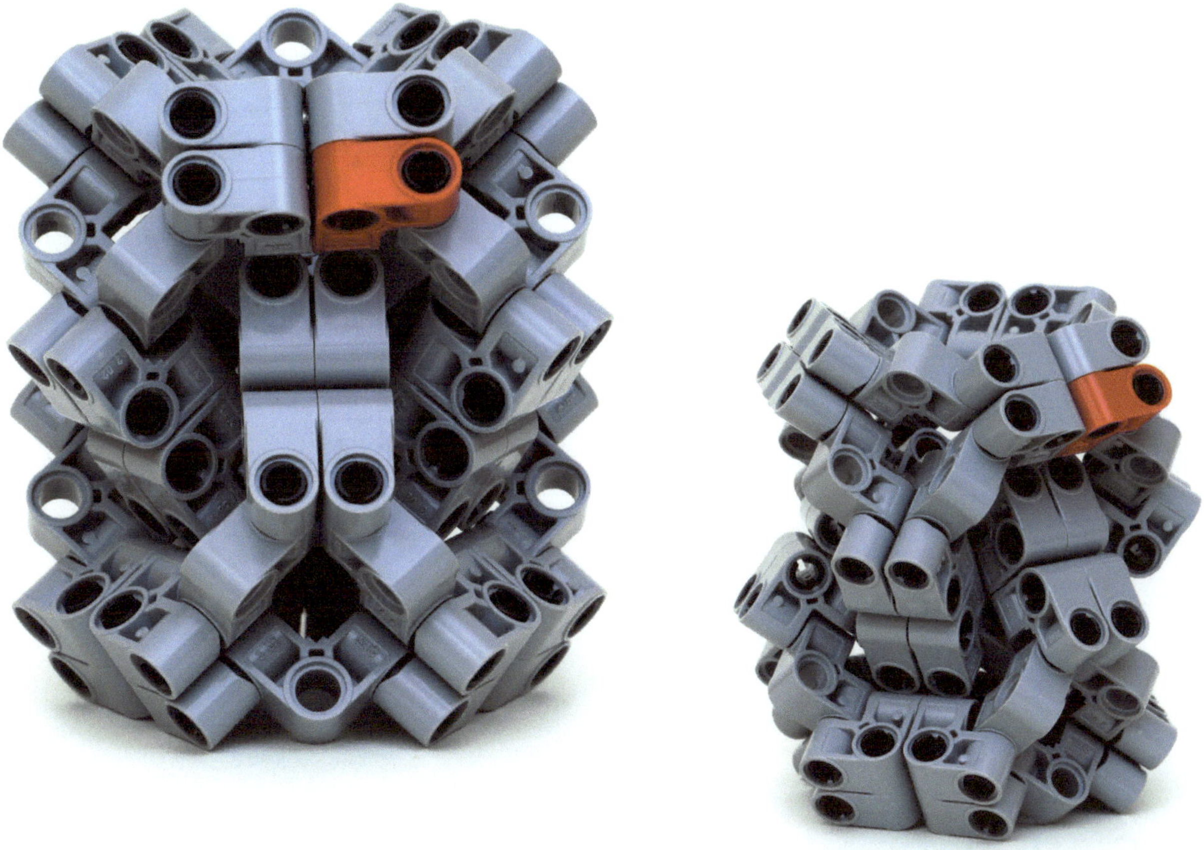

2 Intersecting tetrahedra – 2 views shown.

This is a CAD presentation to show the intersection configuration of the 2 tetrahedra.

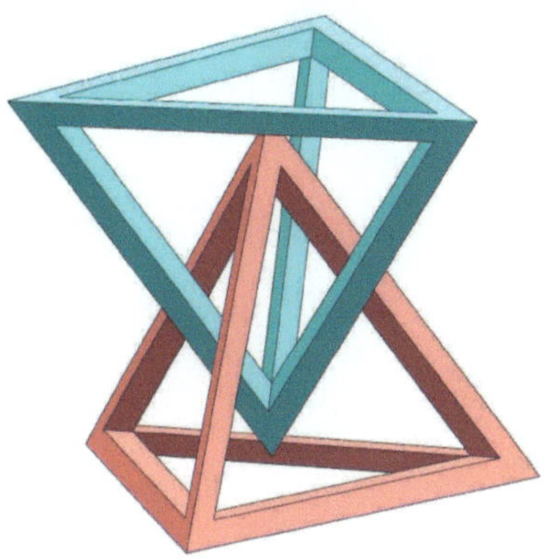

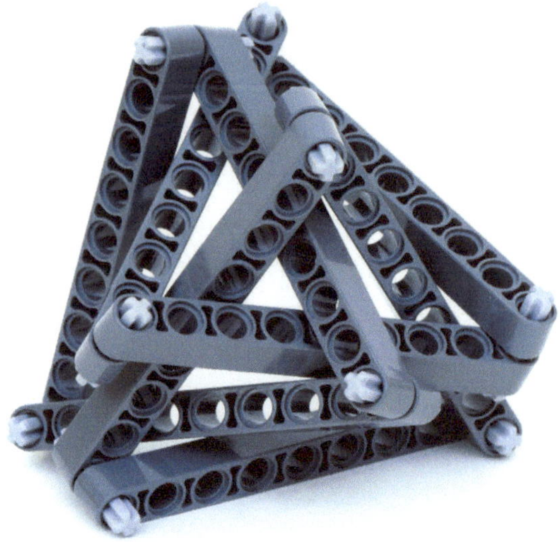

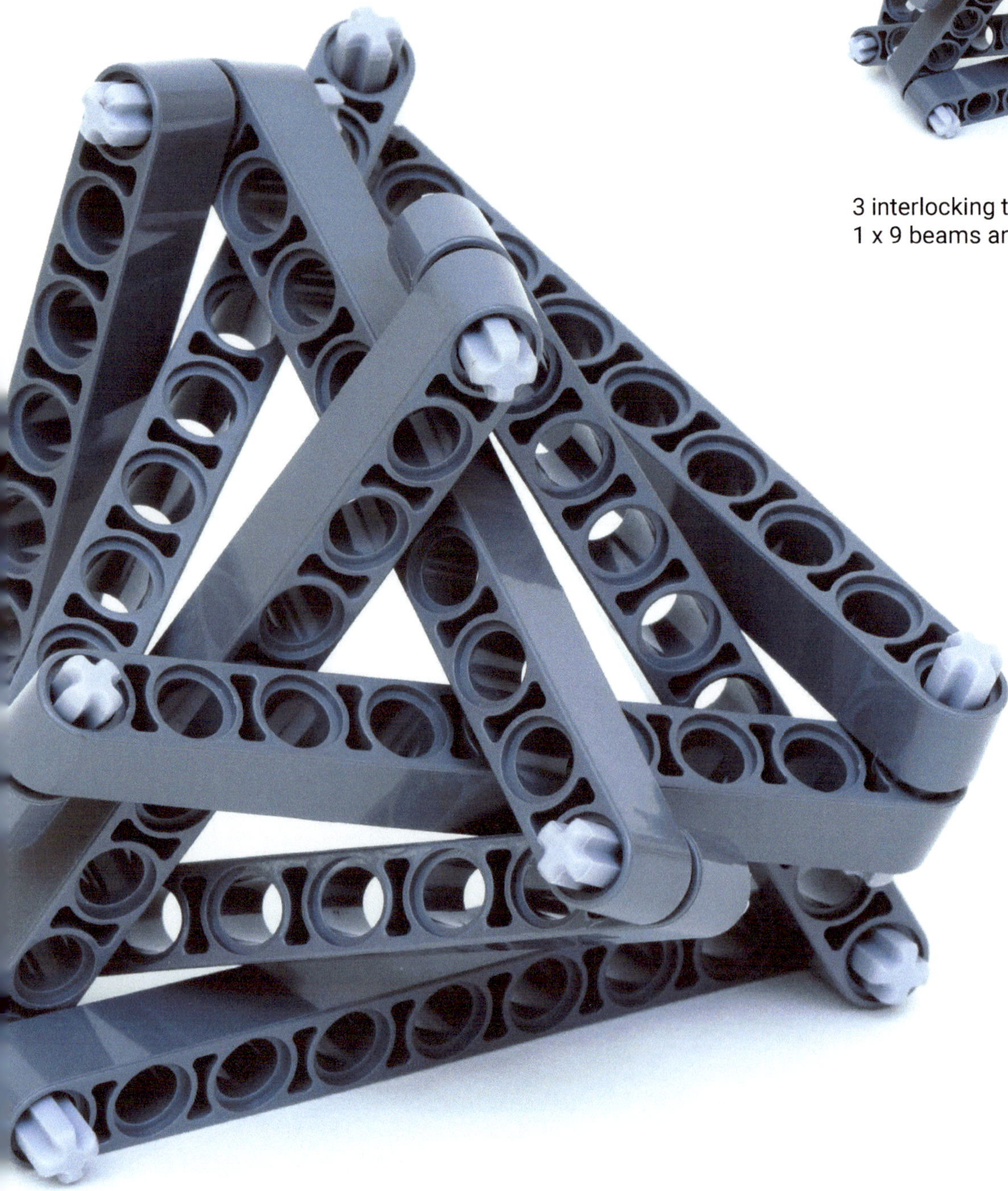

3 interlocking triangles using axle pins as joints.
1 x 9 beams and 3L axles are used.

4 intersecting triangles. The center axis of the delta faces do not meet at a point. The following 2 examples are using this same weave type.

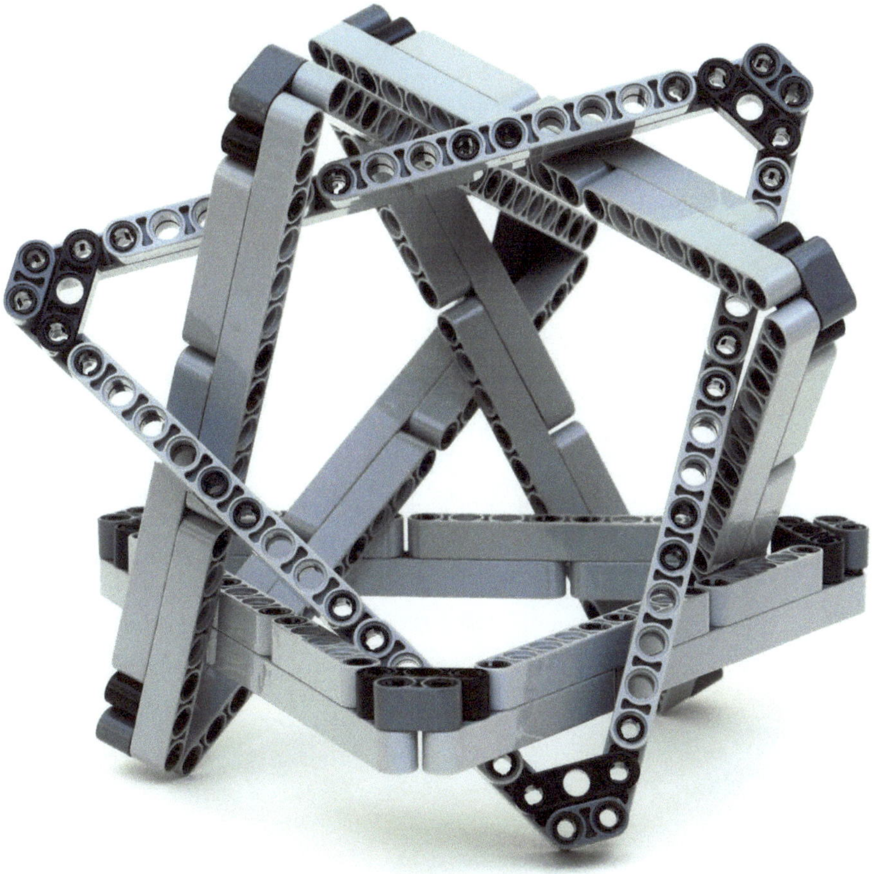

1 edge of the triangle face.

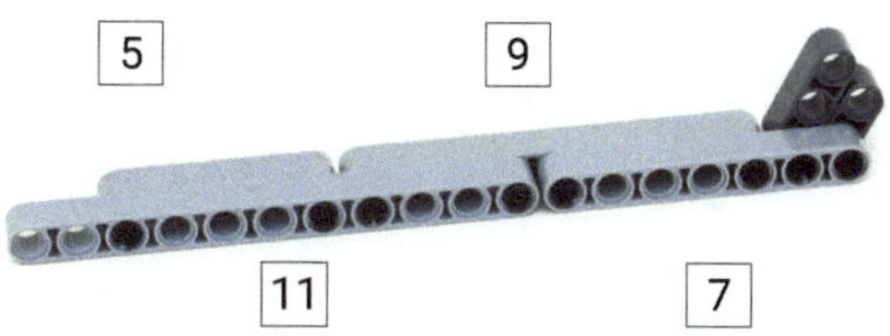

x 12

Parts:

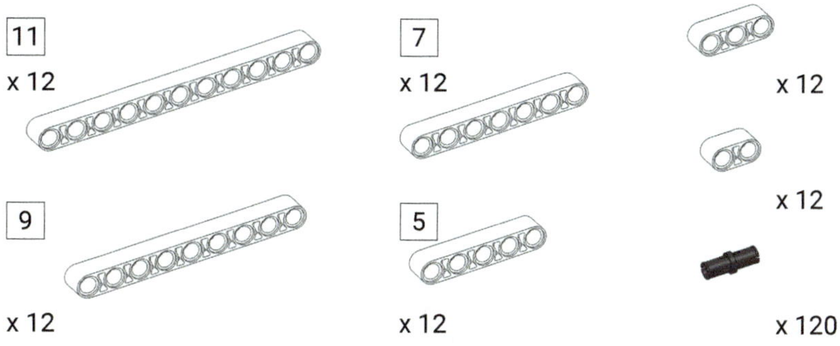

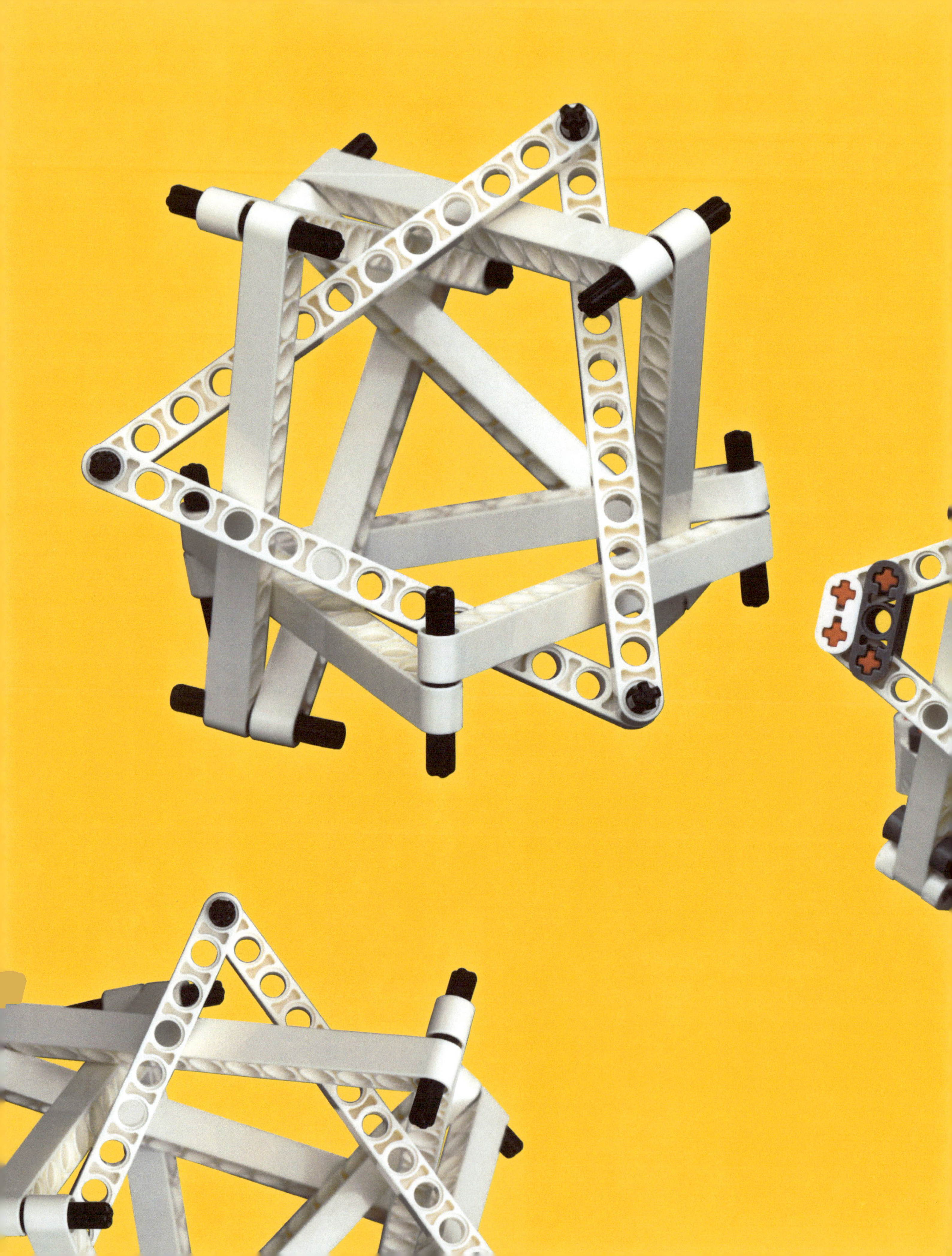

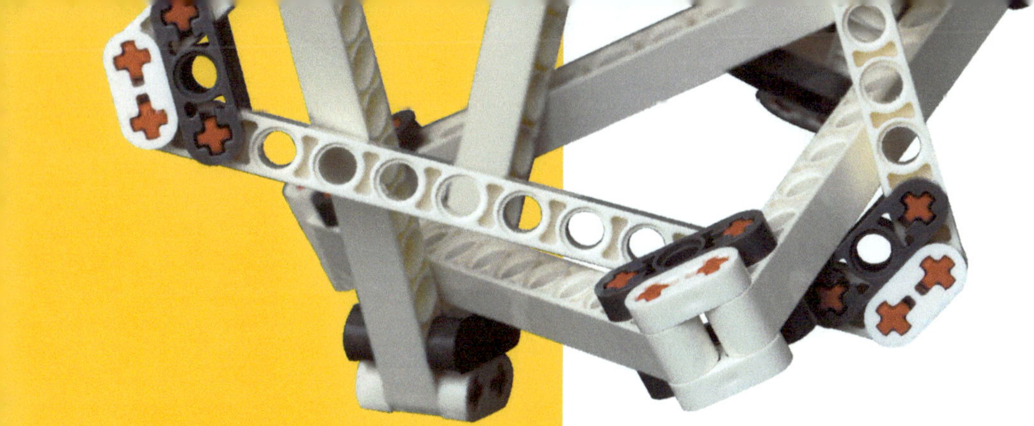

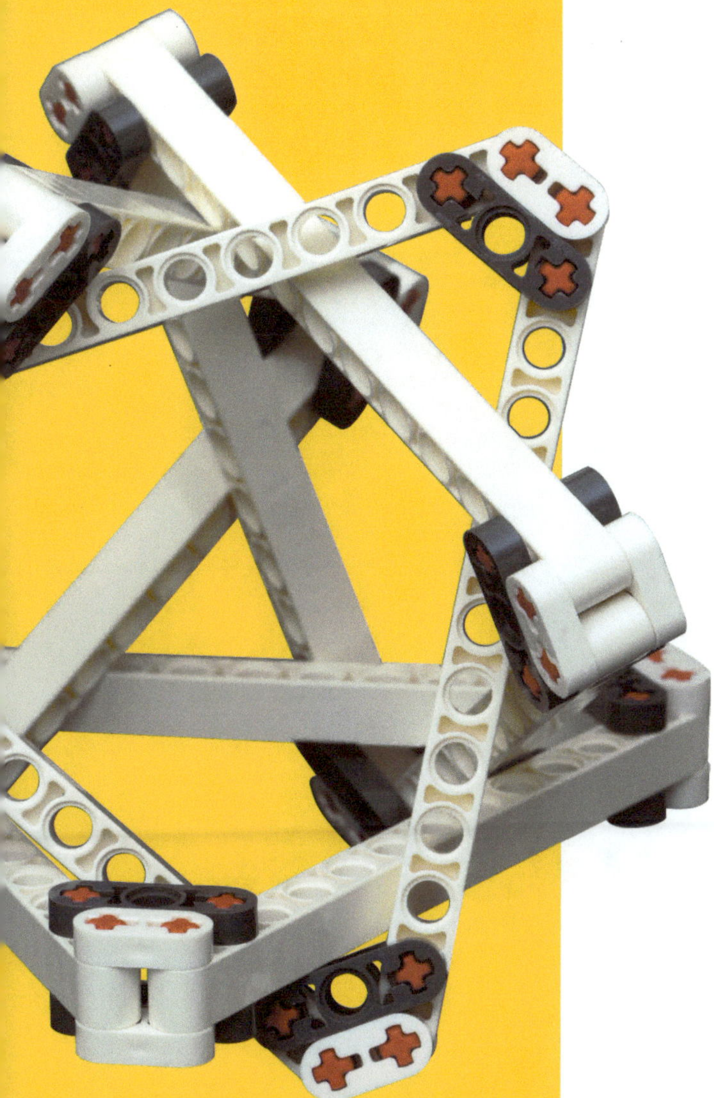

4 intersecting triangles

Parts:

13

x 12

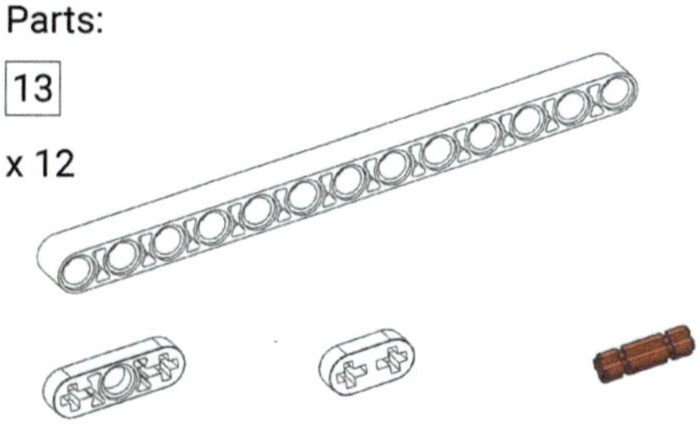

x 24 x 24 x 48

CHAPTER 2

CUBE

The cube is one of the five Platonic solids. There are many ways to make this model in Lego. Here are some construction methods used to form the corners of the cubes. 2 types of corner construction are used – tripod types and hexapod types. Here are a number of examples to show the various possibilities of using different types of connectors. Additional examples are models that are derived from the cube.

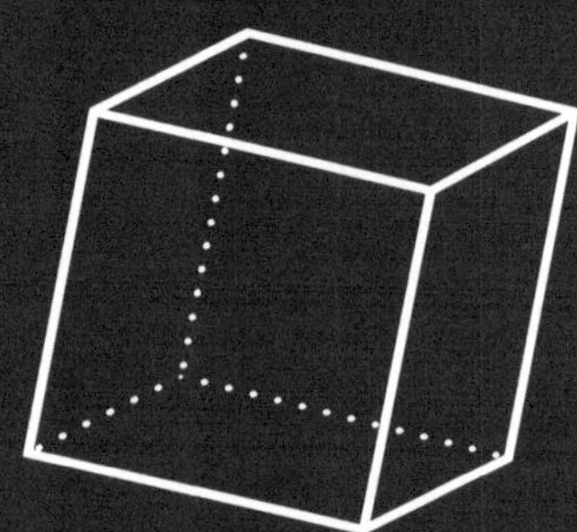

Tripod Types - Cube corner construction

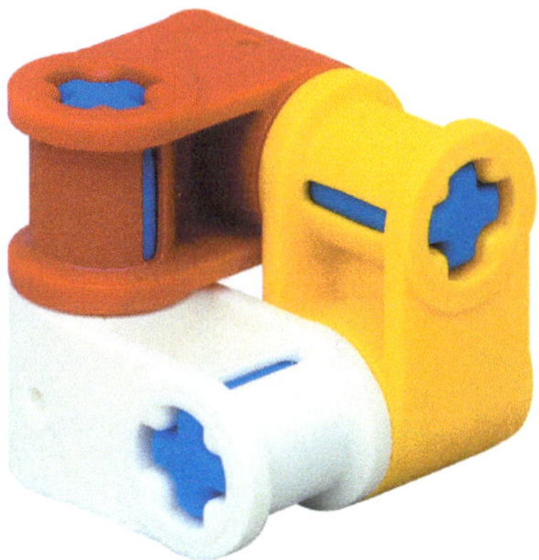

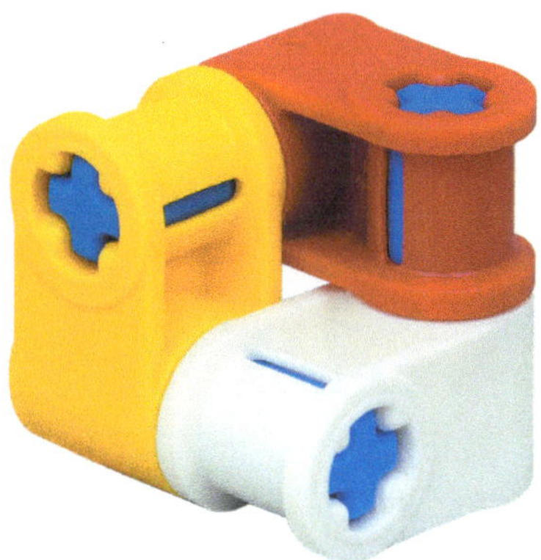

This is a basic tripod type configuration and may need to have extended pins or axles for connecting to each other. Right hand and left hand shown. They are a mirror image of each other.

Assemble the parts in a partial 'loose' fit then gradually pressed in simultaneously.

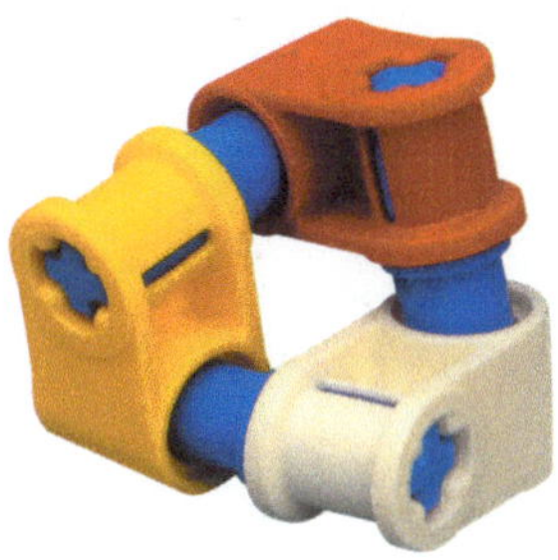

Parts:

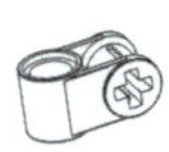

x 3

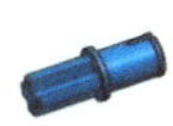

x 3

A cube using the above corner type, 4 right hand and 4 left hand corners.

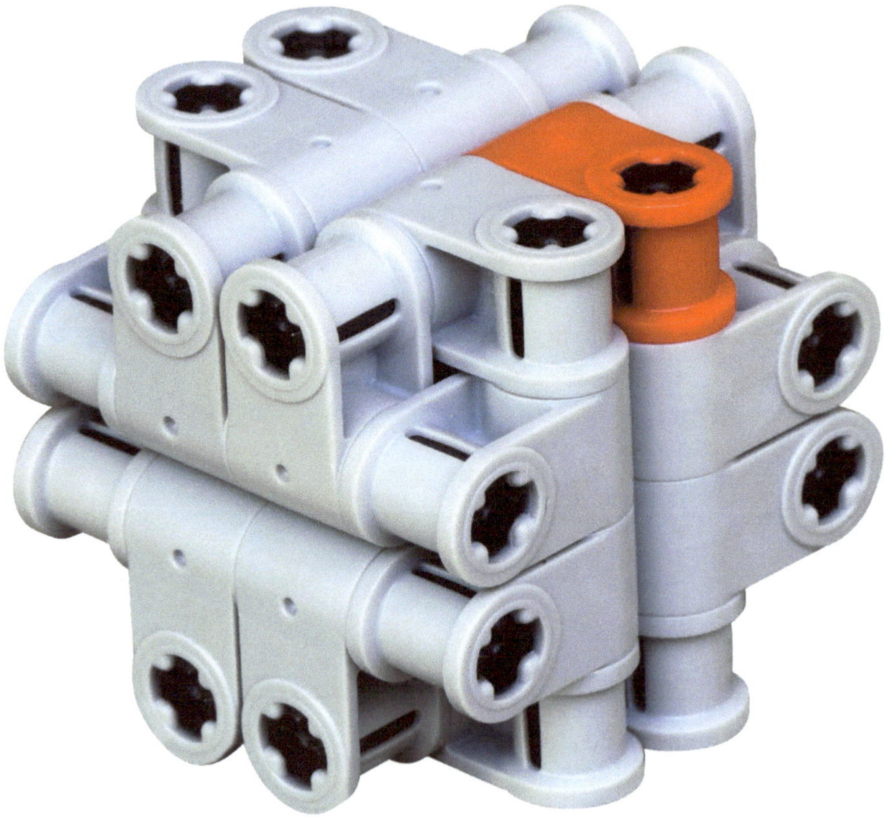

Parts:

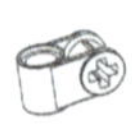

x 24

4

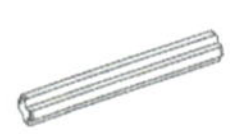

x 12

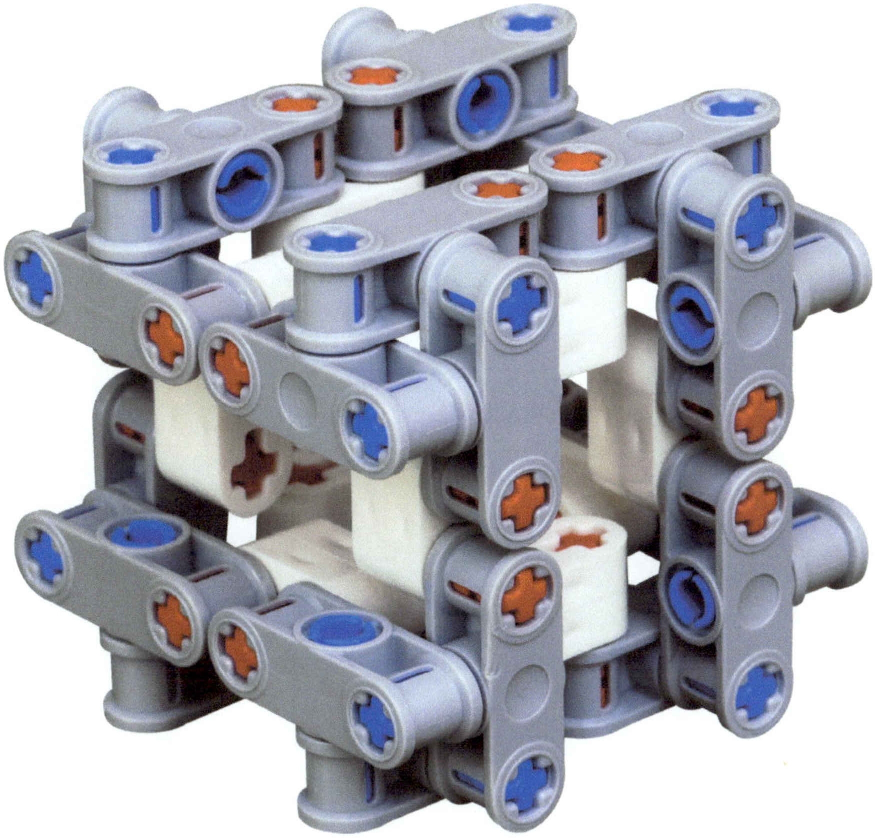

Right hand and left hand corners.

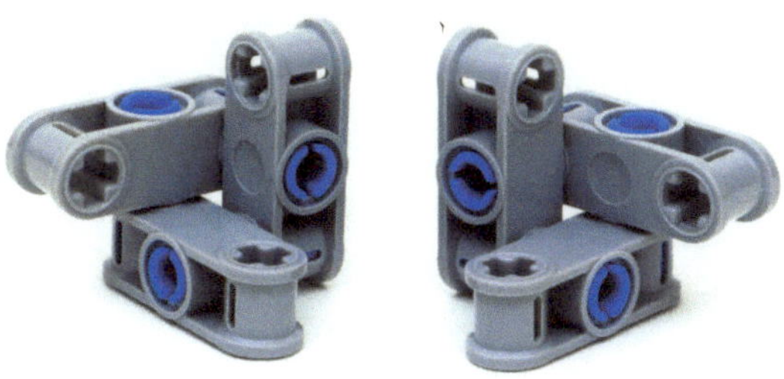

Parts:

x 24 x 24 x 24 x 24

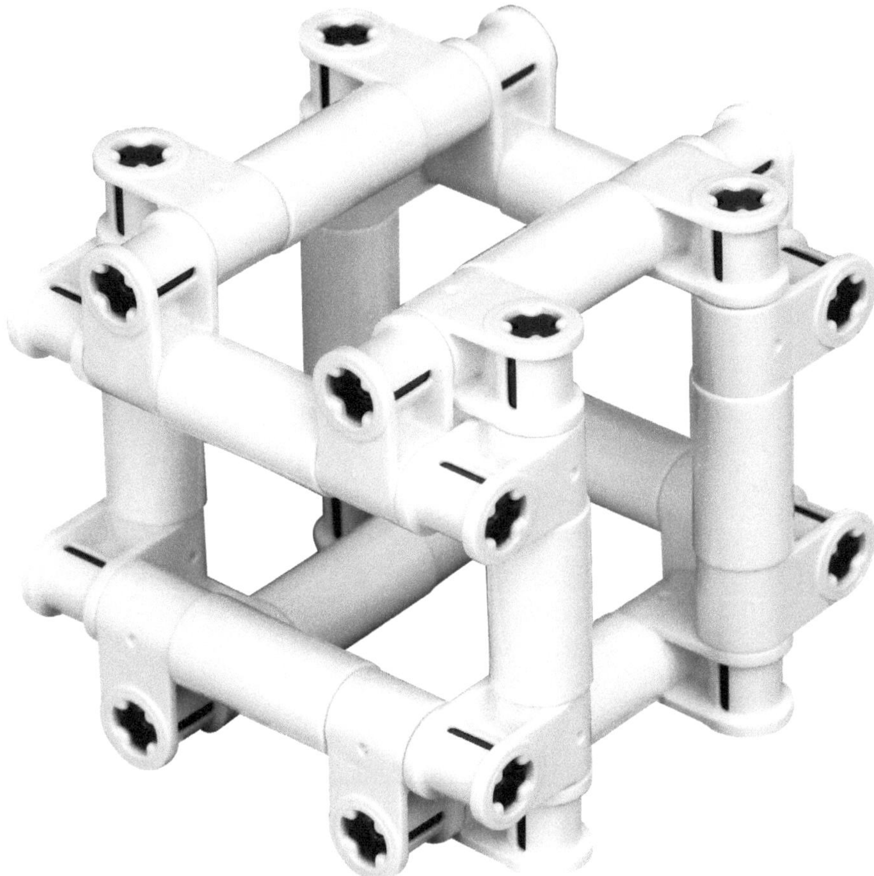

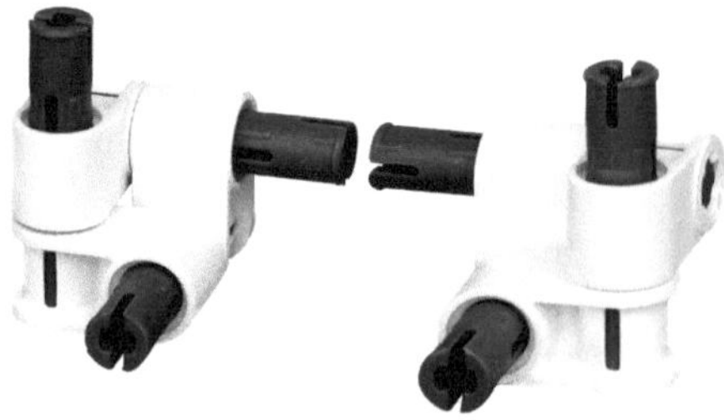

Right hand and left hand corners of the above cube. A complete cube will have 4 right hand and 4 left hand corners. This will be the same requirement for all the examples of cube corner construction.

Parts:

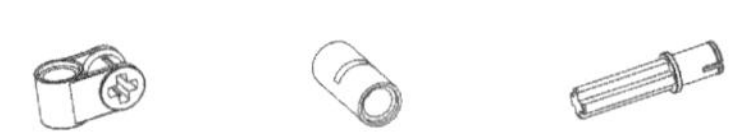

x 24	x 12	x 24

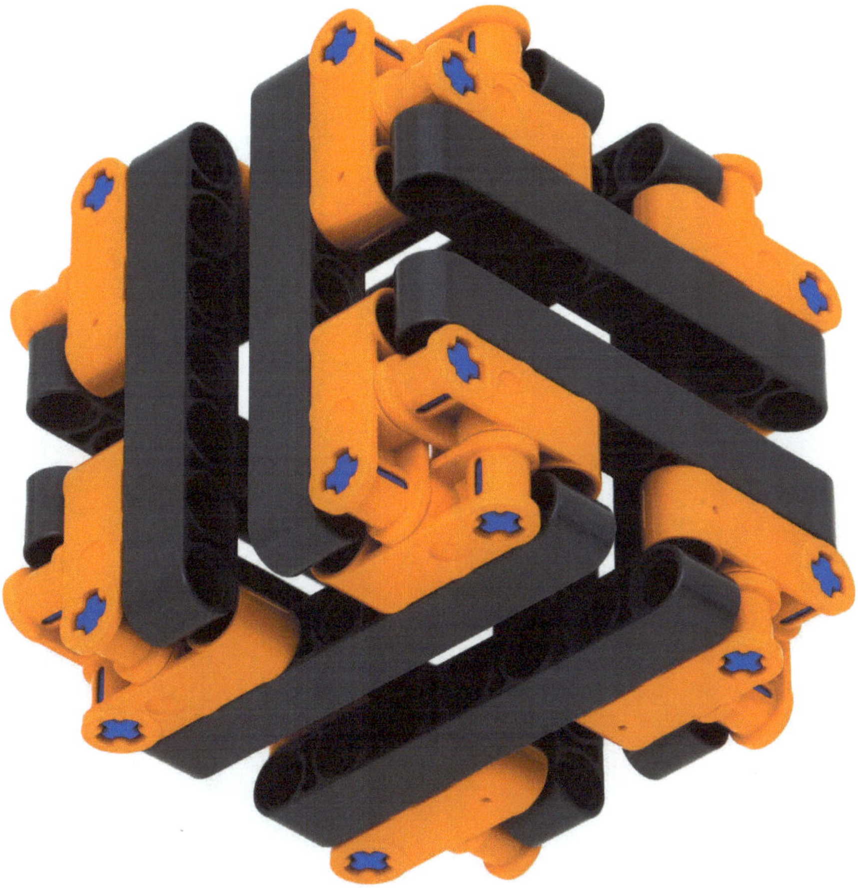

Parts:

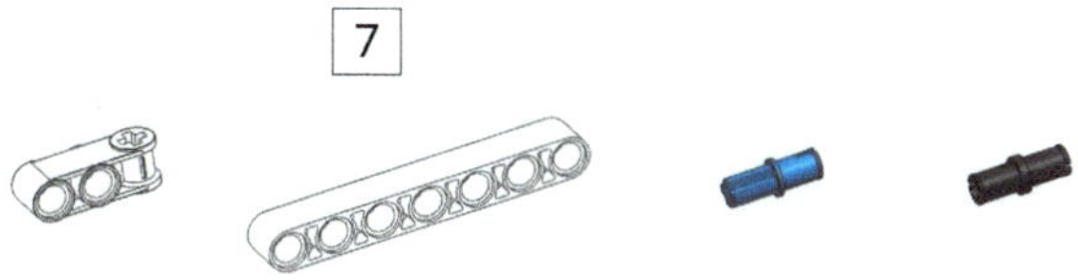

x 24 x 12 x 24 x 24

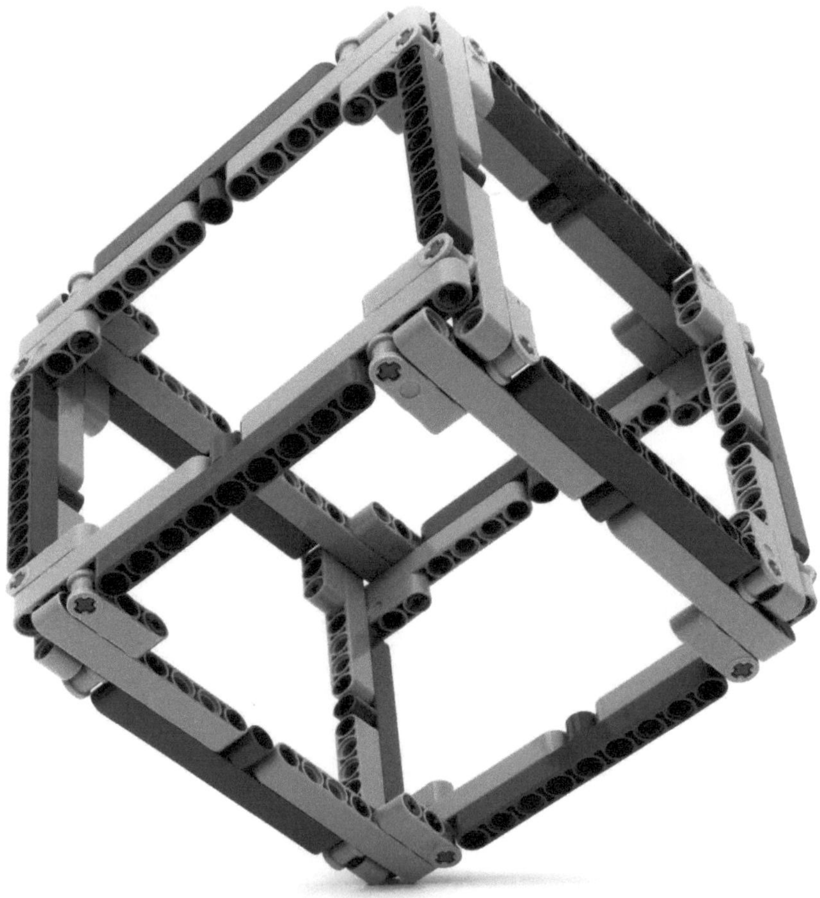

Cube corners.

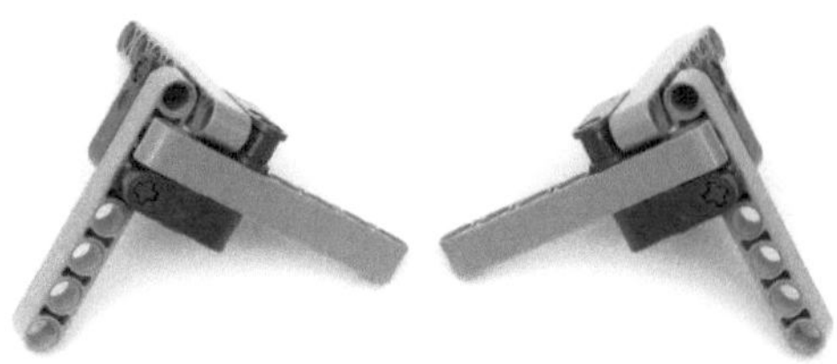

Parts:

9 x 12

7 x 24

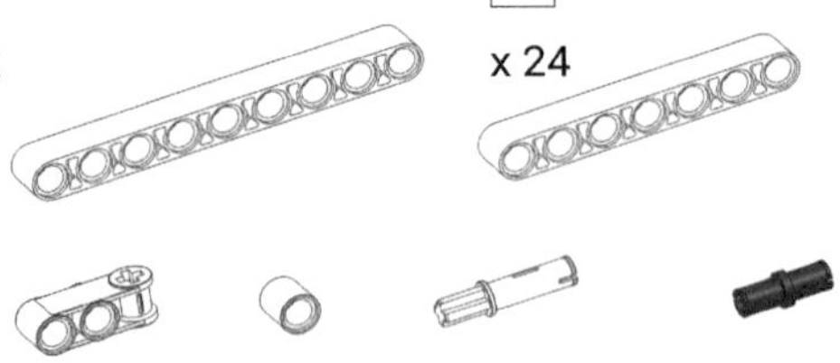

x 24 x 12 x 24 x 36

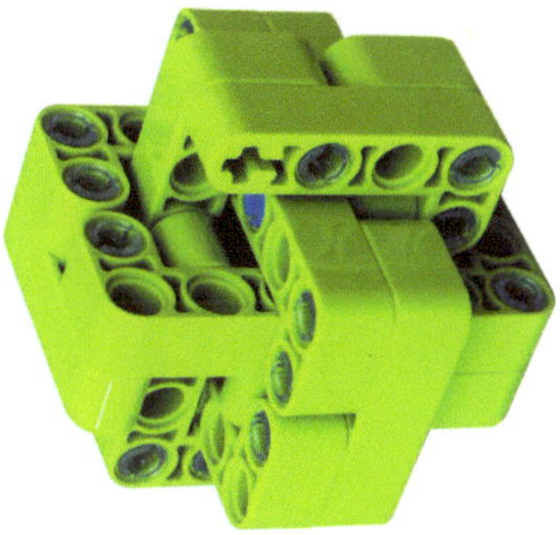

3 interlocking rectangles in a Borromean link configuration.

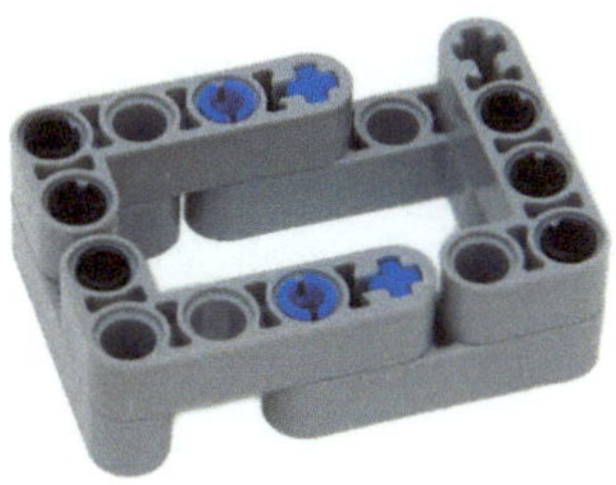

Rectangle construction. **x 3**

Parts:

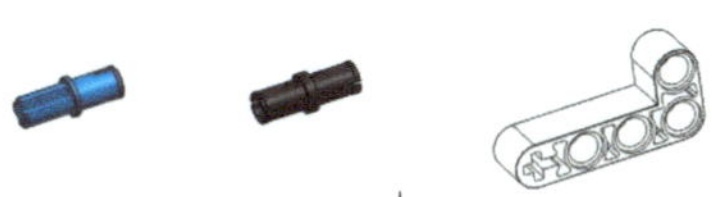

x 12 x 18 x 18

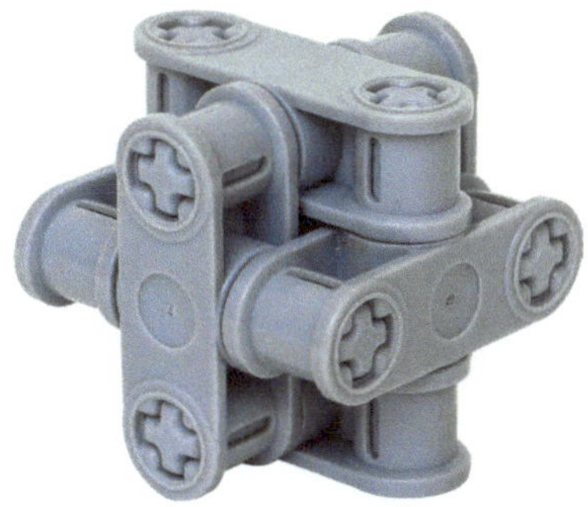

6 interlocking connectors in a tripod type configuration.

Parts:

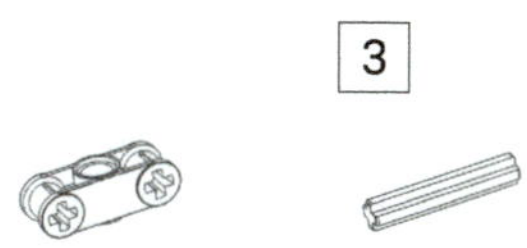

x 6 x 6

3 pairs of connectors assembled orthogonally to each other to form the corner.

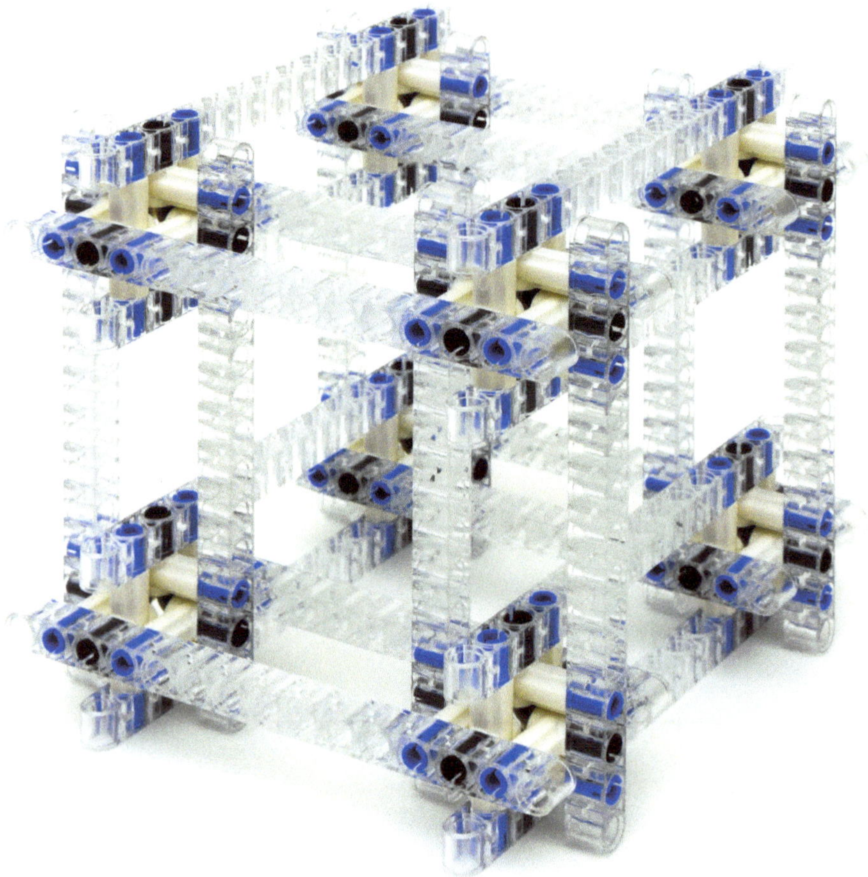

Parts:

x 96 15

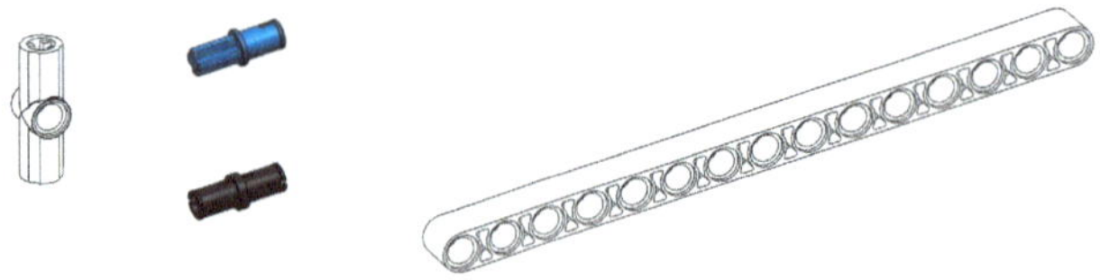

x 48 x 48 x 24

A cube with each corner formed by 3 rectangles.

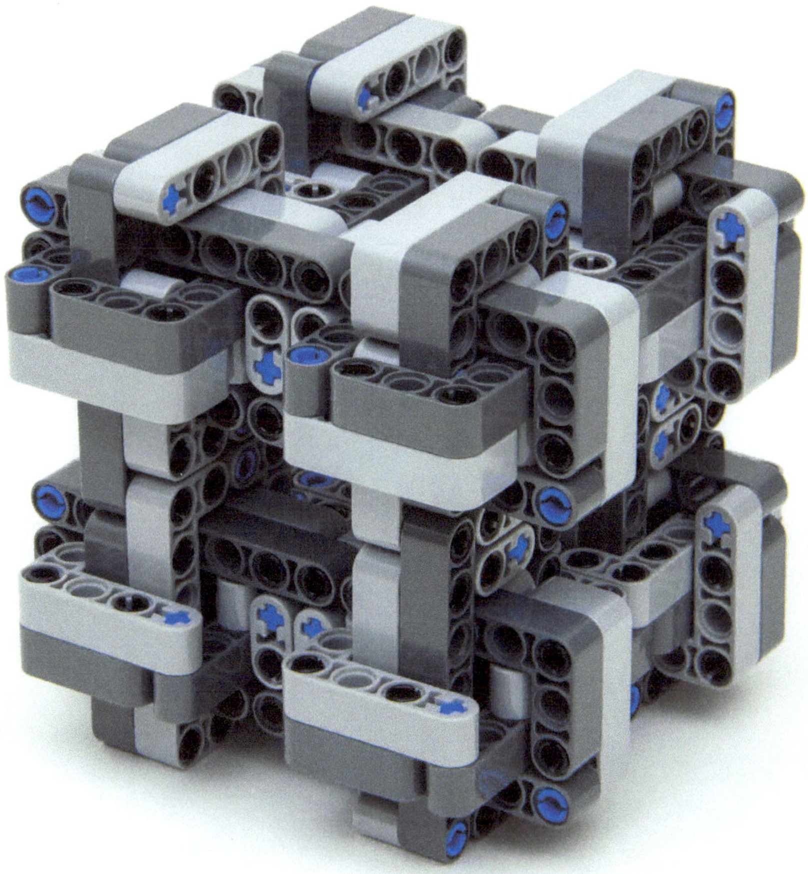

Top and bottom views of 1 rectangle.

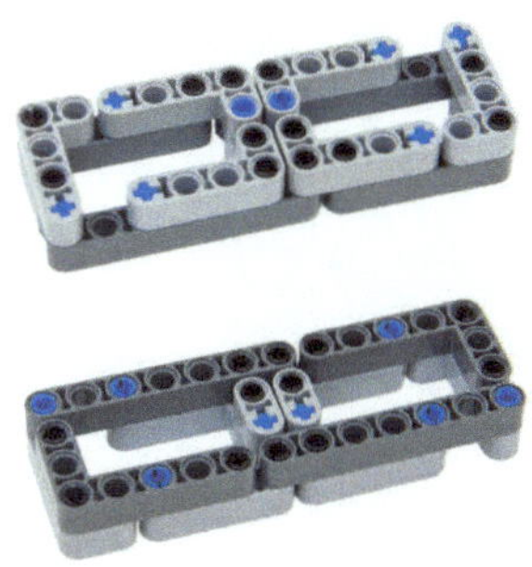

Parts:

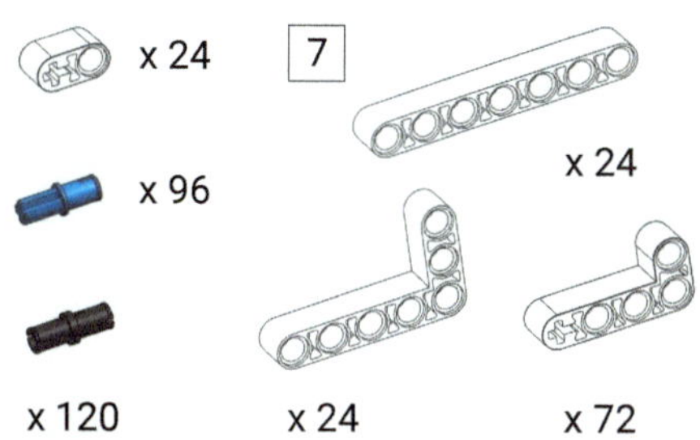

HEXAPOD TYPES
Cube corner construction

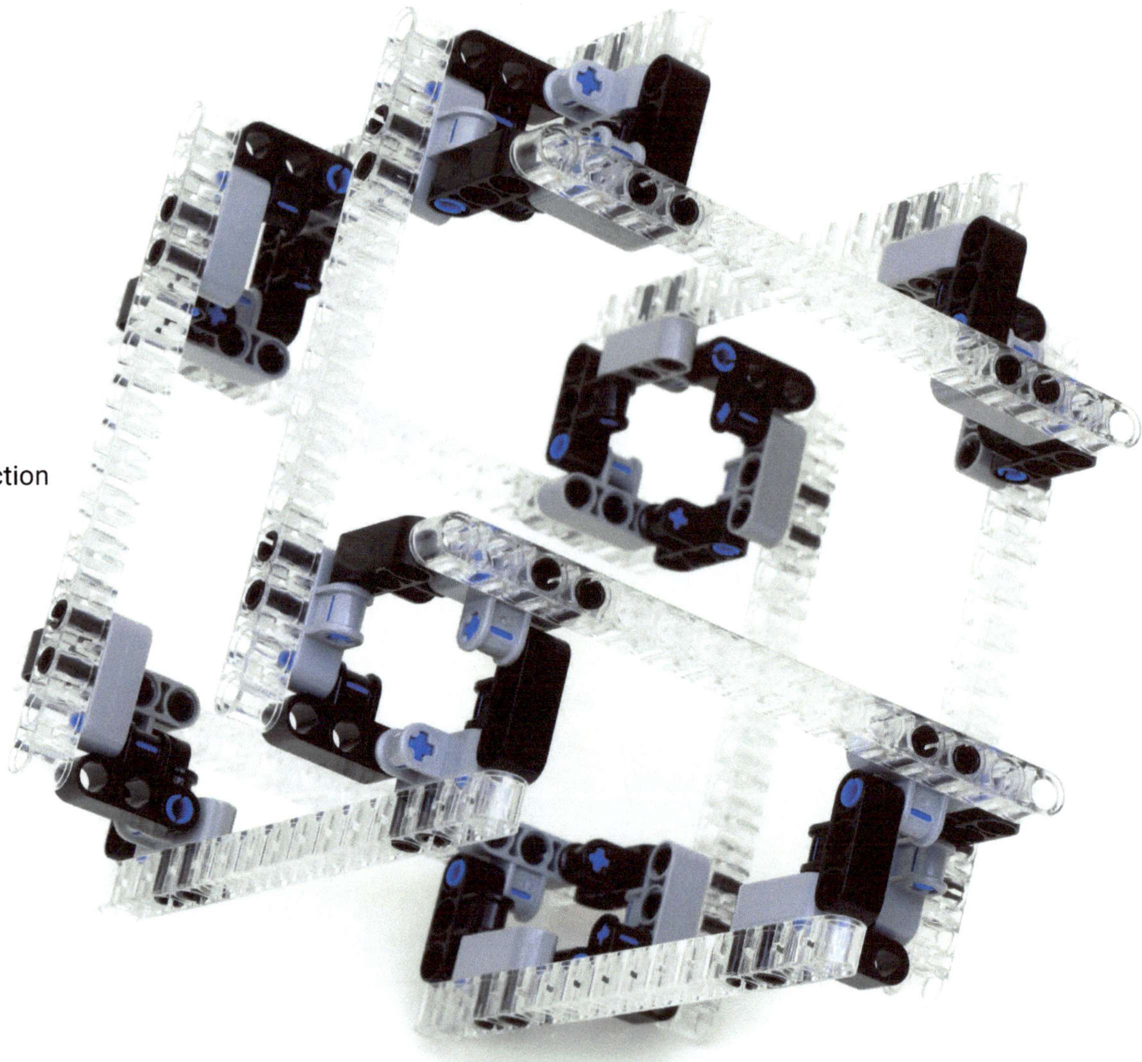

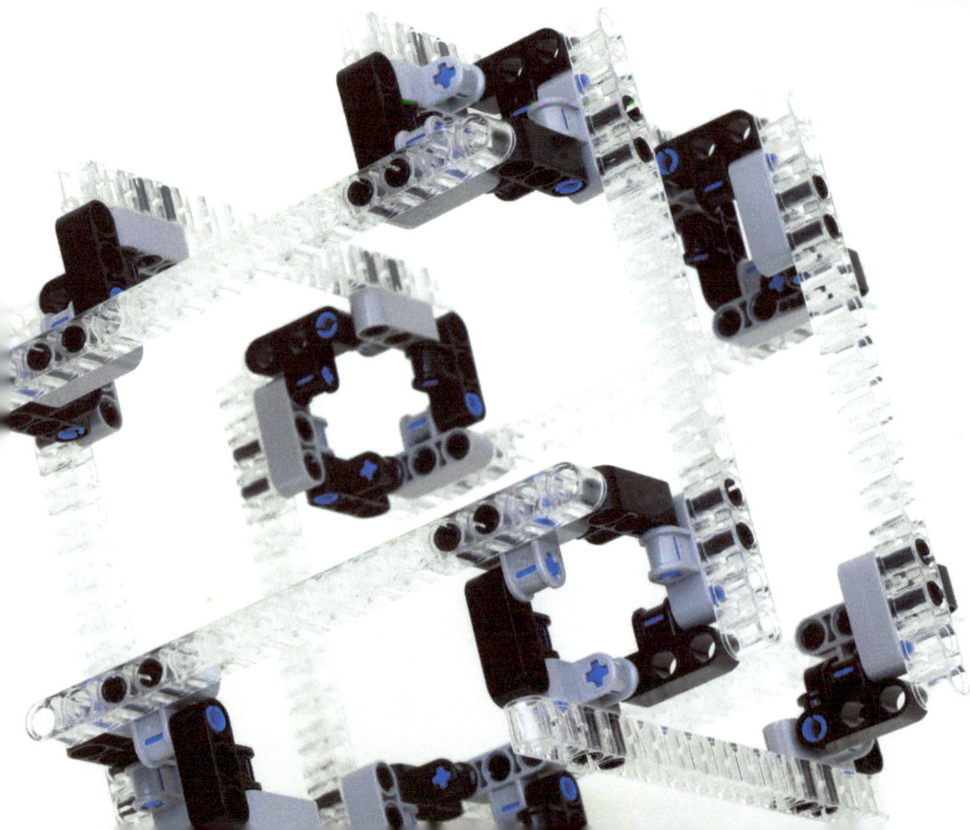

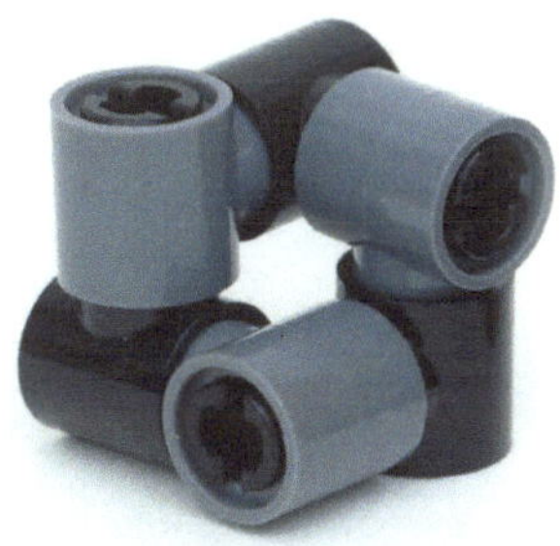

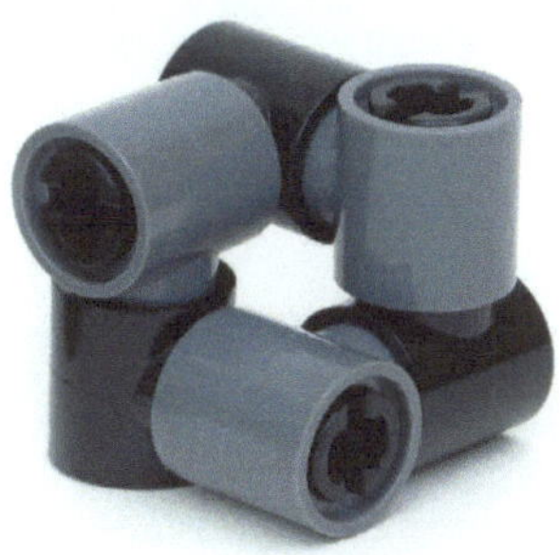

This is a basic hexapod type concept used as a corner construction. Right hand and left hand shown.

Parts:

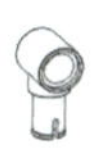

x 12

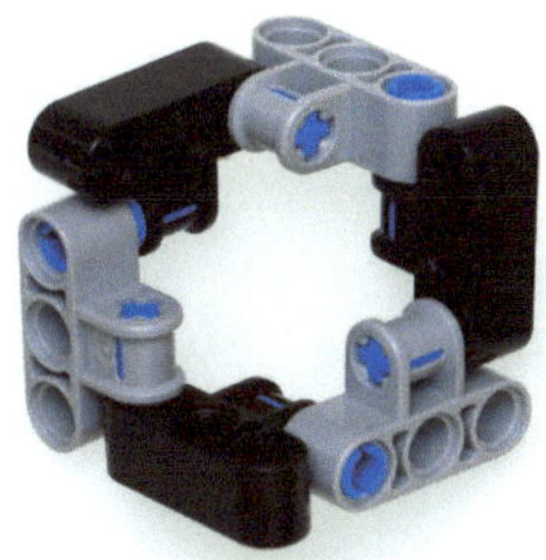

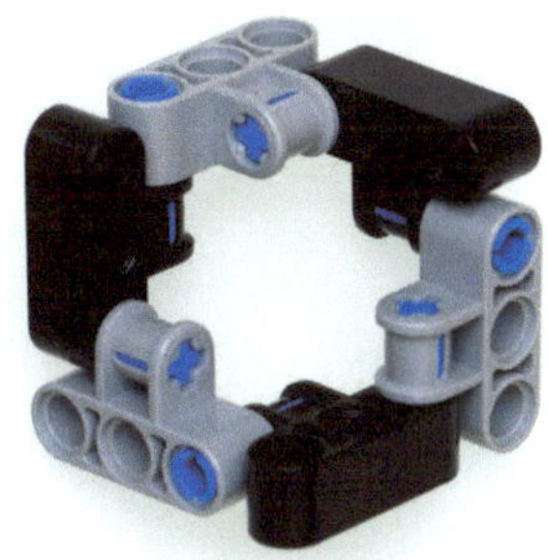

Hexapod type corners, right and left hand.

Parts:

15

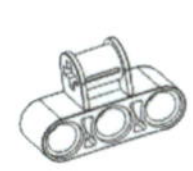

x 48

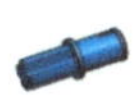

x 48

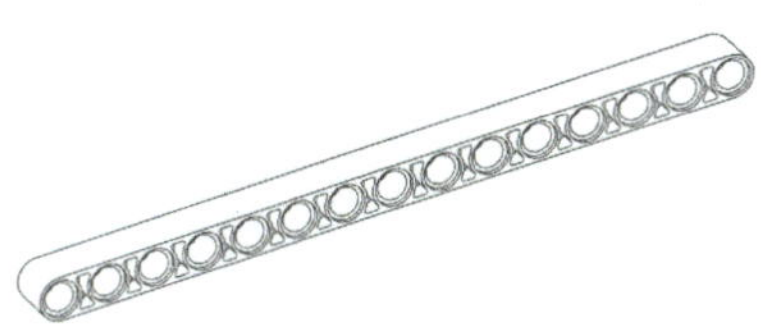

x 12

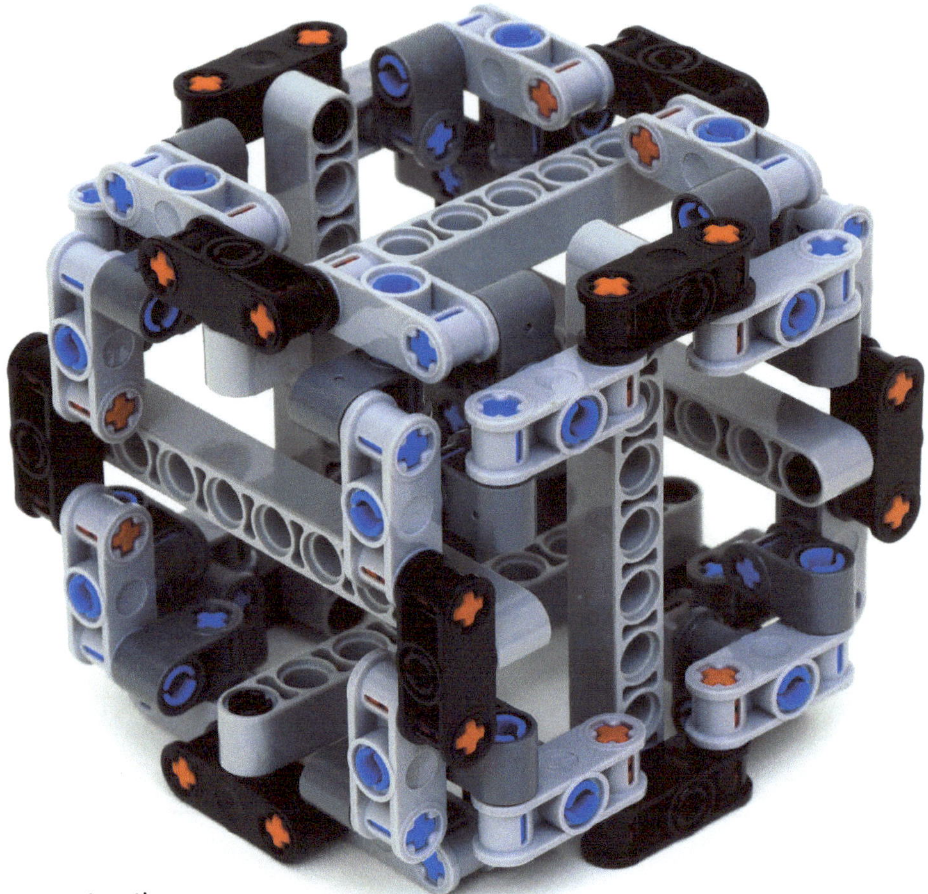

Hexapod type cube corner construction used for the above cube. 4 right hand and 4 left hand corners are required for each cube in these examples.

Hexapod type corner.

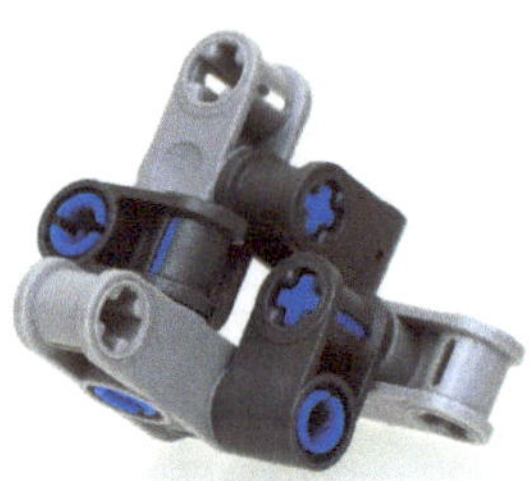

Parts:

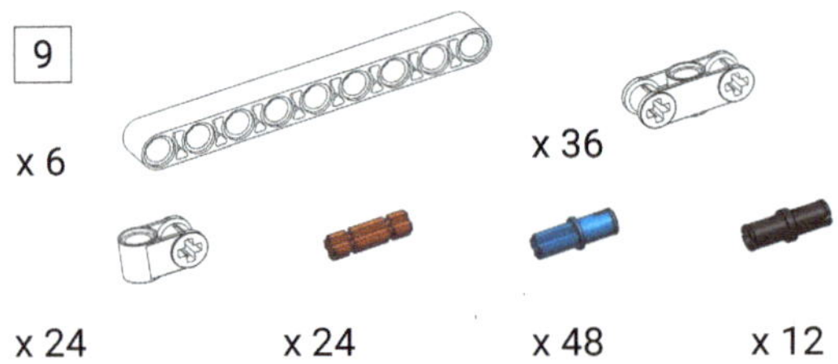

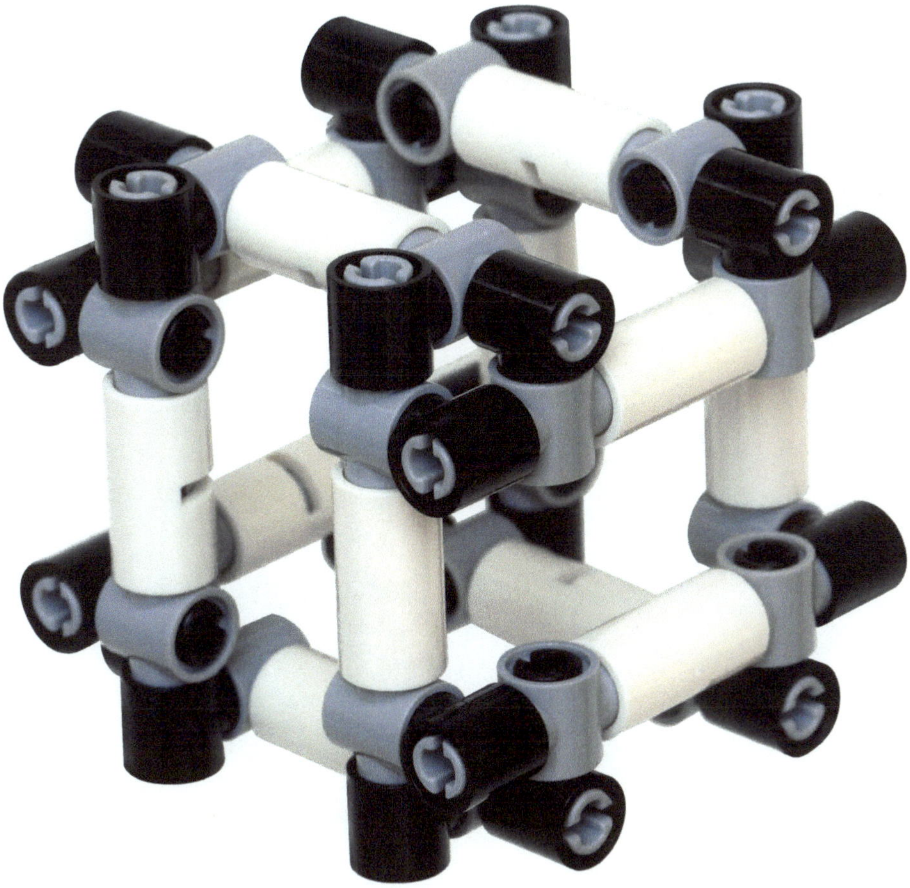

Hexapod type corner.

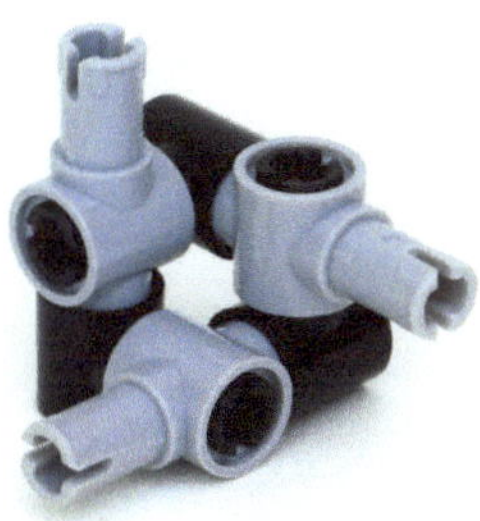

Parts:

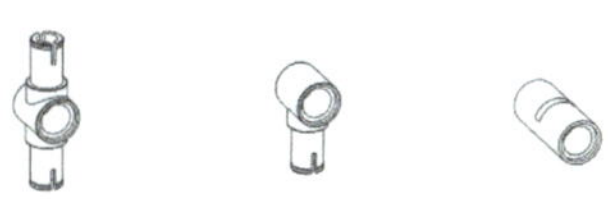

x 24 x 24 x 12

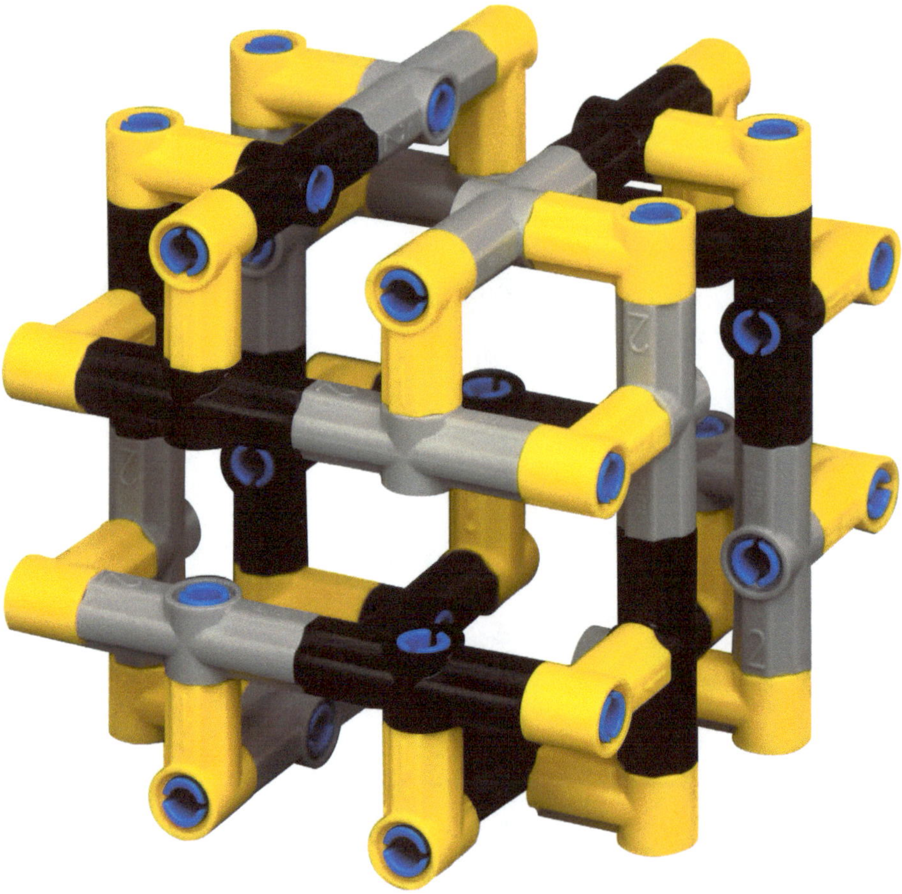

Parts:

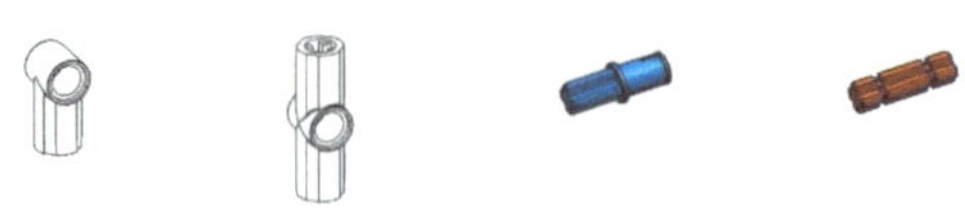

x 24 x 24 x 48 x 12

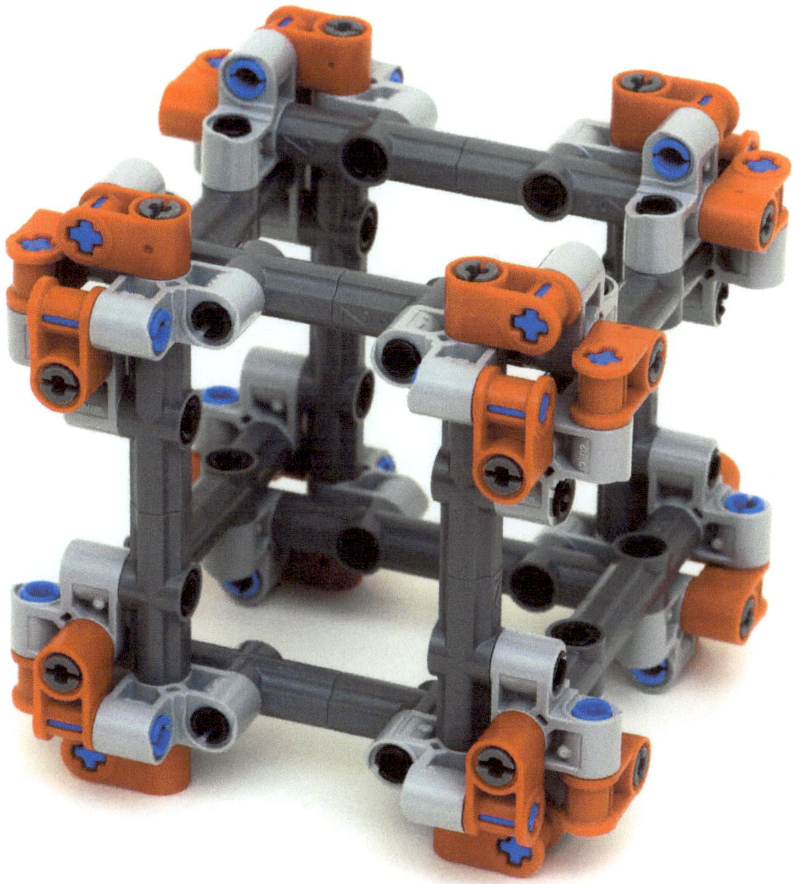

Parts:

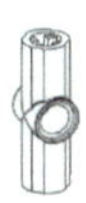 x 24

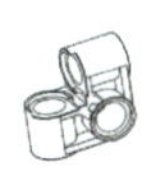 x 24

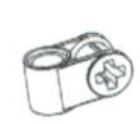 x 24

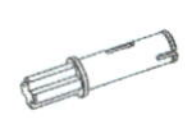 x 24

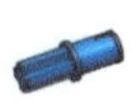 x 24

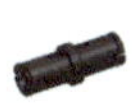 x 24

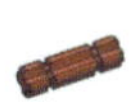 x 12

CUBE STRUCTURES
Models derived from the cube

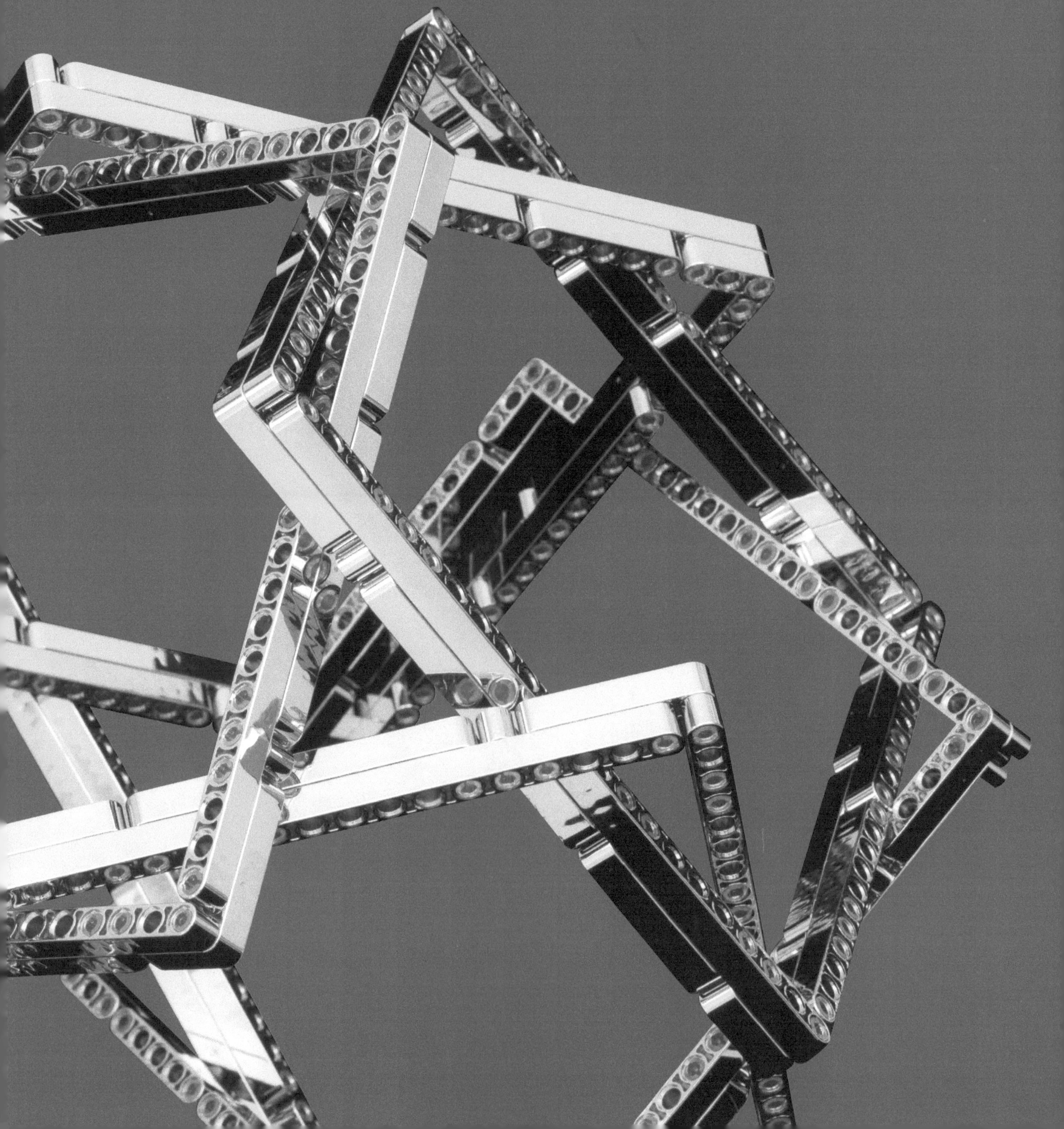

This is an 'inverse' version of the 6 intersecting squares with an external weave configuration. In general, for the construction of intersecting face polygons, there are generally 2 solutions as shown in the examples of this section.

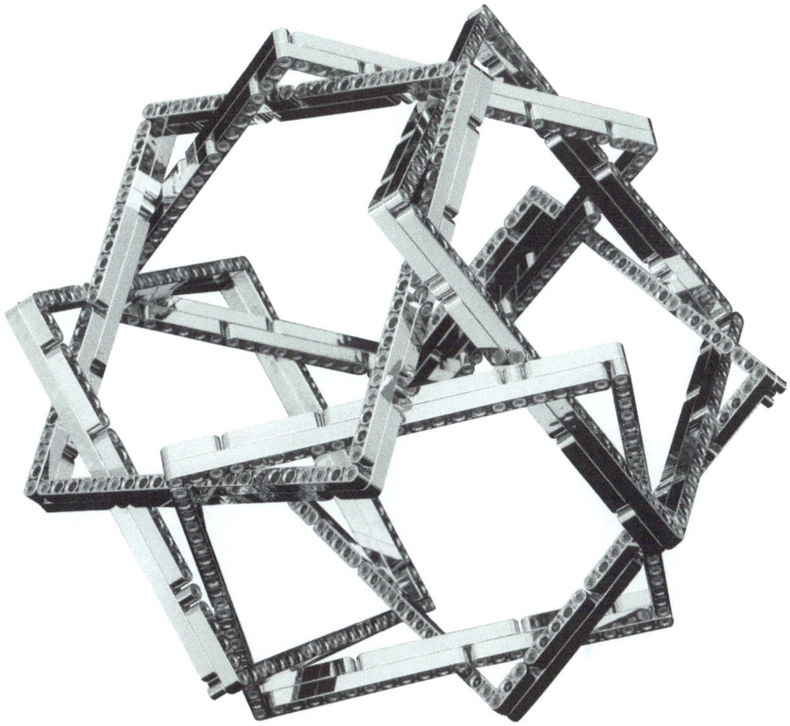

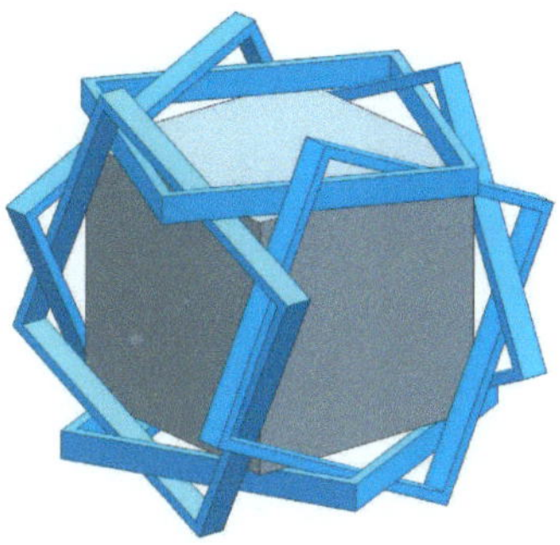

Download
https://grabcad.com/library/6-squares-geometric-sculpture-1

x 24

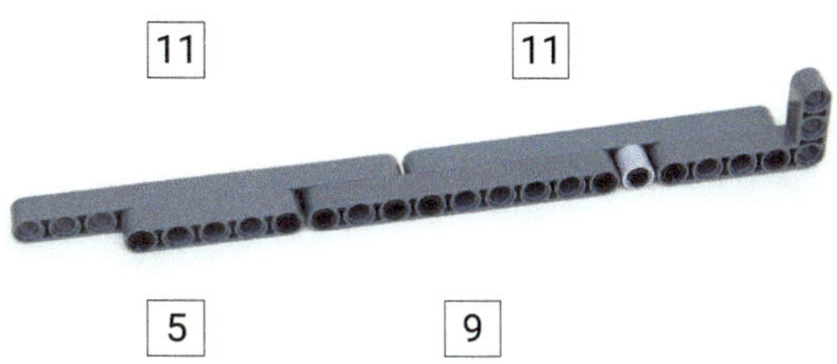

Parts:

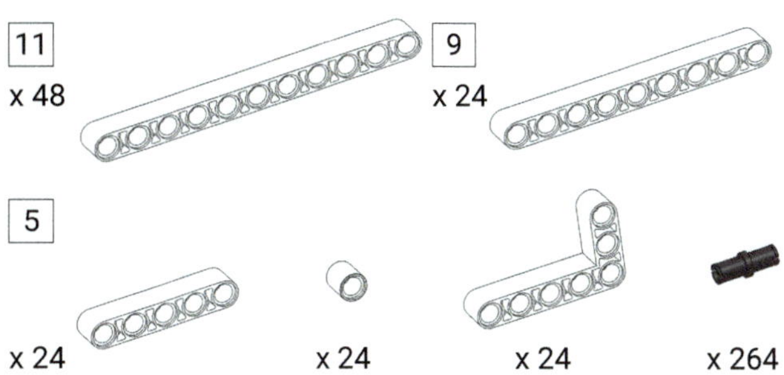

6 intersecting squares.

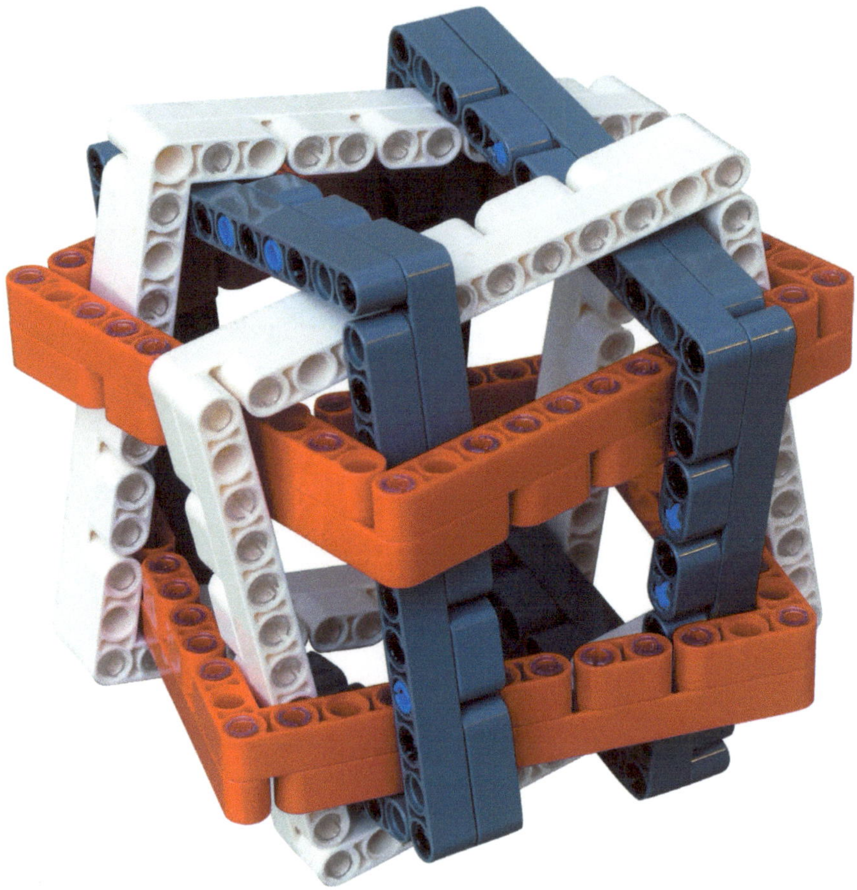

x 24

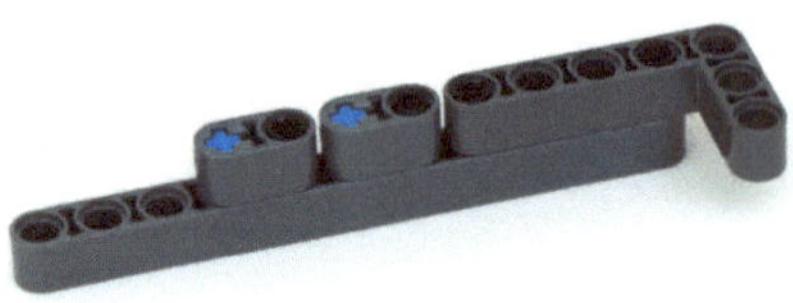

Parts:

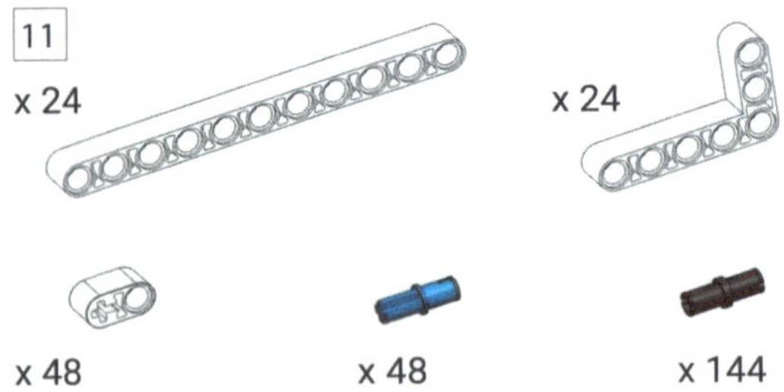

6 intersecting squares using connectors.

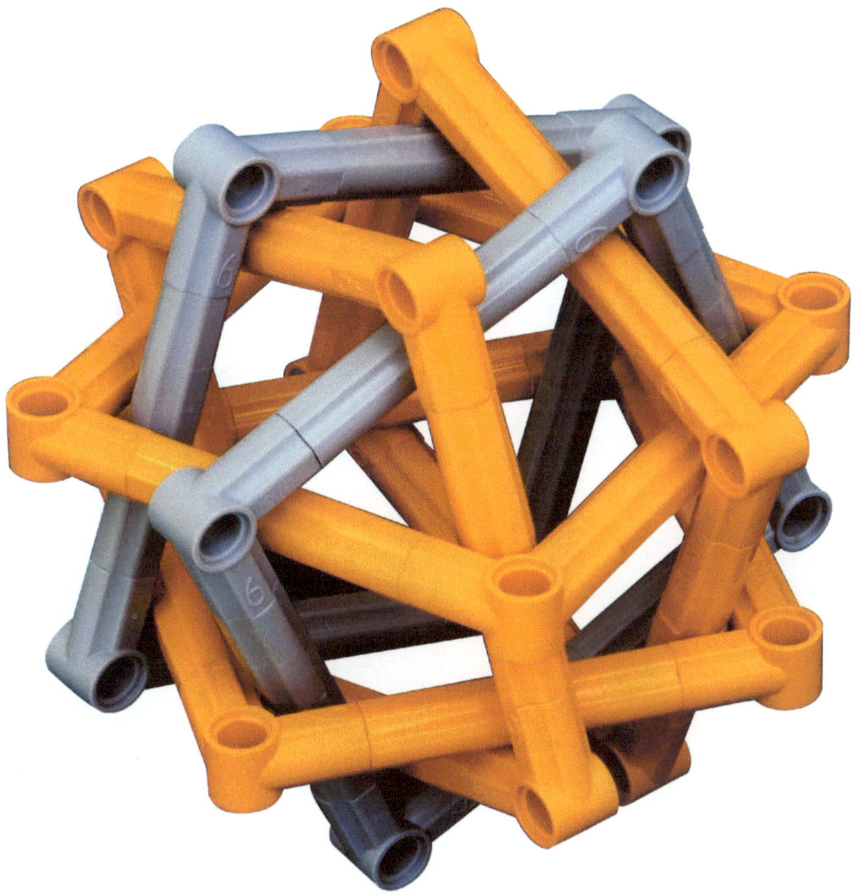

Parts:

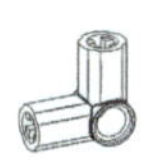	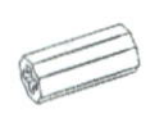	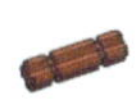
x 24	x 48	x 72

OCTAHEDRON

An octahedron is a polyhedron with 8 faces, 12 edges and 6 vertices.

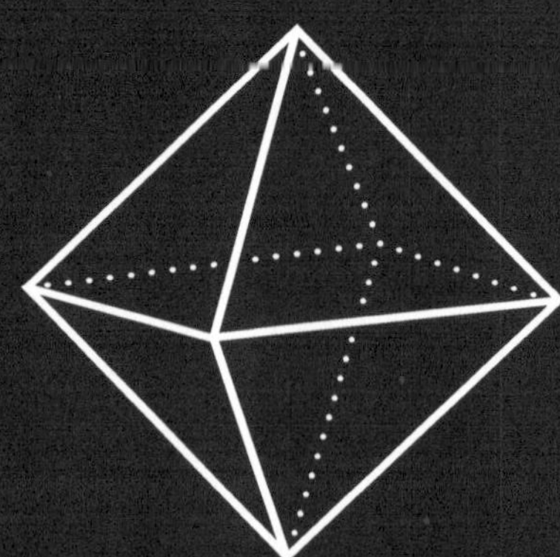

CHAPTER 3

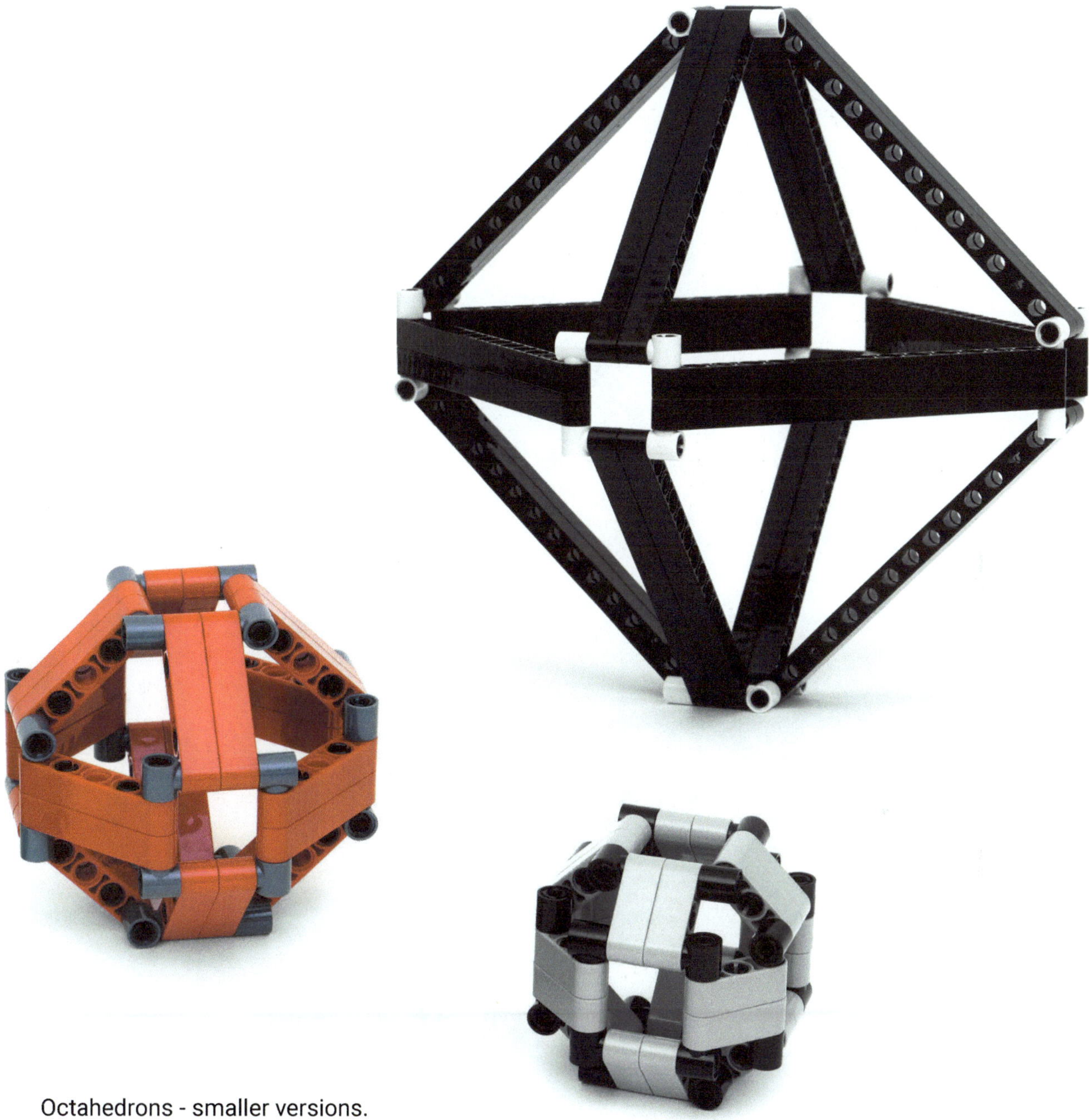

Octahedrons - smaller versions.

Parts:

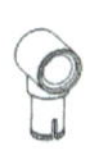

x 24

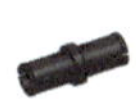

x 48

15

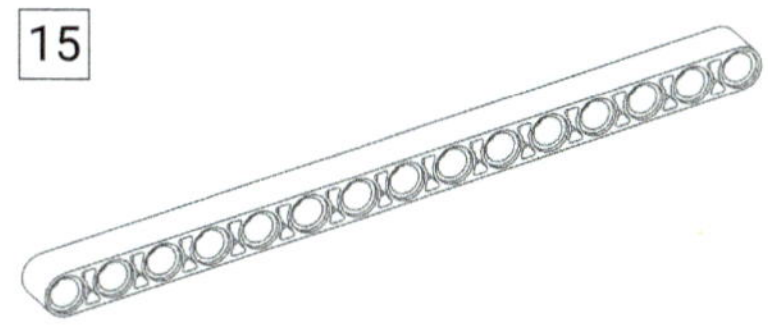

x 24

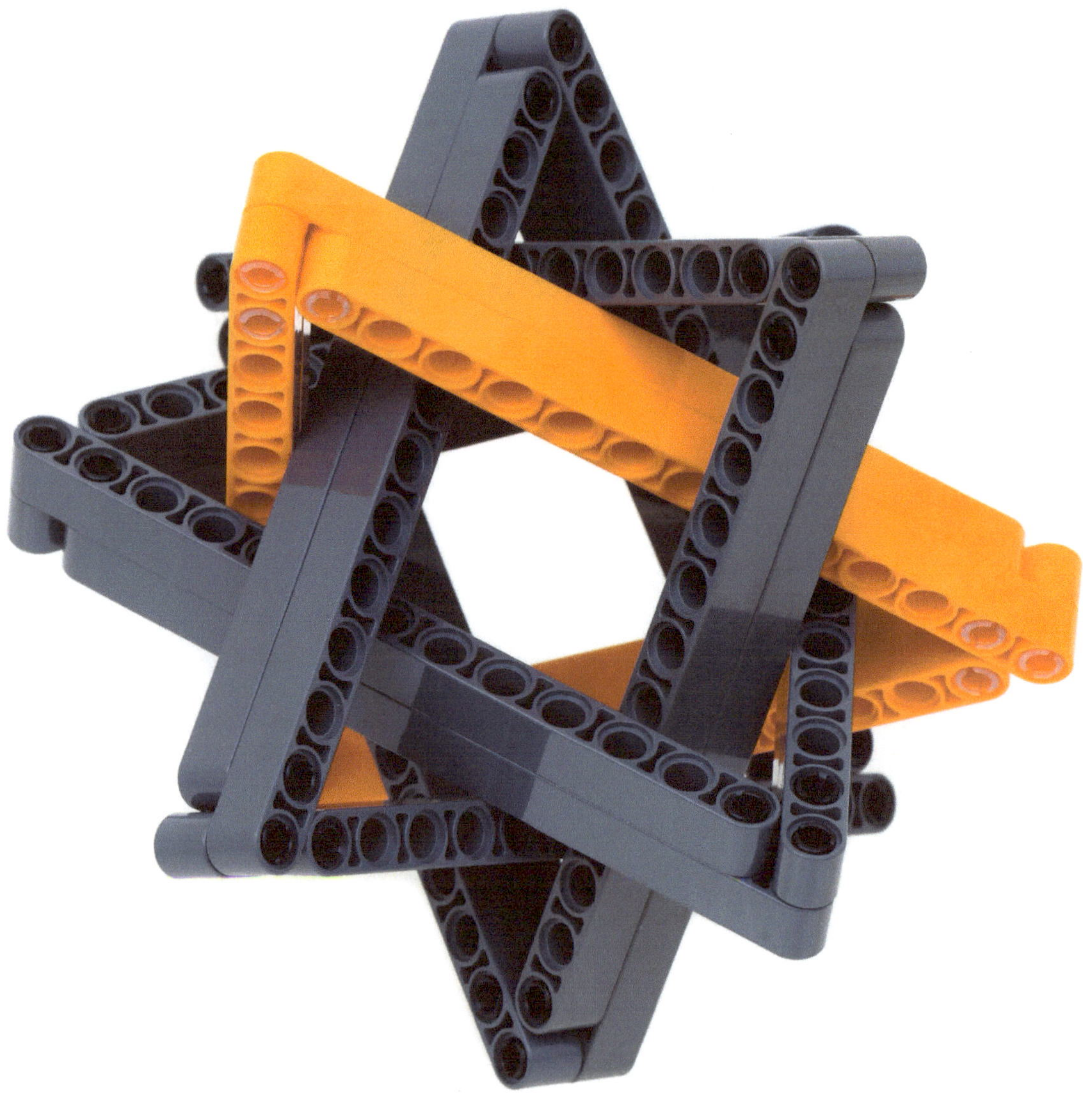

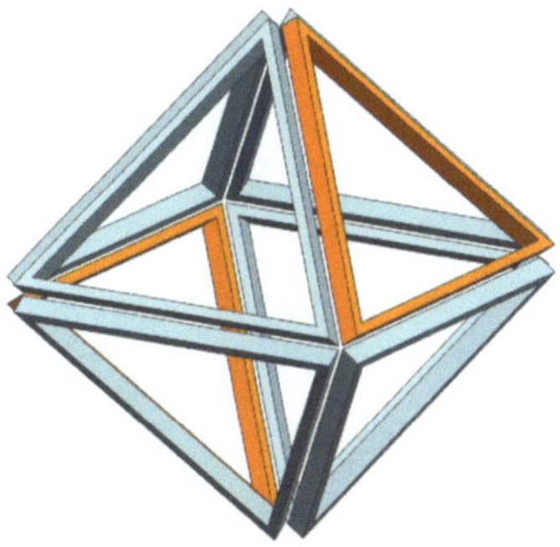

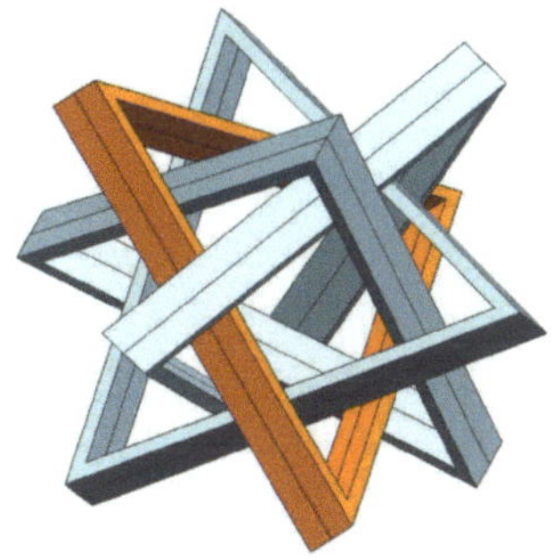

4 Intersecting Triangles.

The 4 intersecting triangles sculpture is derived from an octahedron by rotating and translating opposite faces simultaneously to meet at the center gives a symmetric structure of 4 triangles. This structure can also be derived from a tetrahedron.

To assist in the assemblies, I have links to download the CAD models which I had created. I used colored origami paper to help to differentiate the parts and plastic-coated wires to hold the assembly during the work in progress.

The 3D CAD file in "step" format (.stp) file can be downloaded from this link at GrabCad. You can use the 3D viewer available on this site. The CAD image can be manipulated using rotation, zoom and pan. The individual components of the part file can be made visible or invisible as this will help on the building progress of the sculpture. Alternatively, you can use a free CAD software like Fusion 360 and then download the "step" file for viewing.

Download
https://grabcad.com/library/4-triangles-geometric-sculpture-1

Parts:

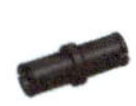

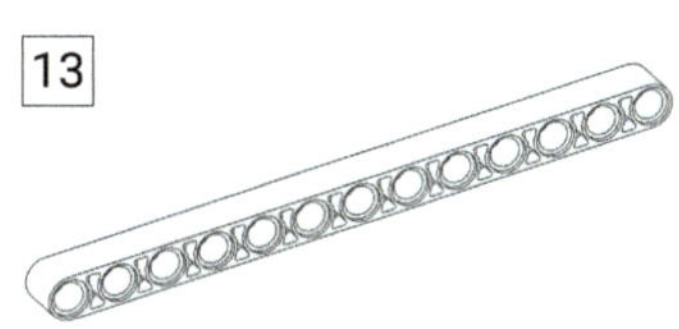

x 36 x 24

8 intersecting triangles.
This is based on the octahedron.

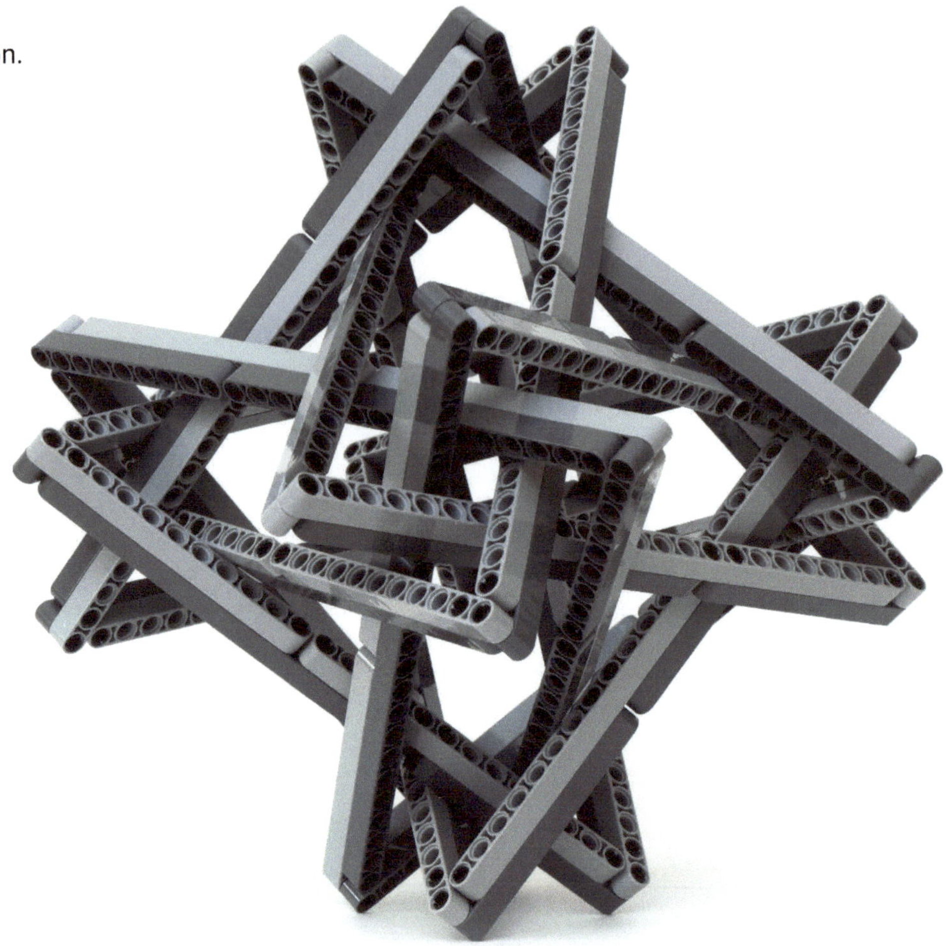

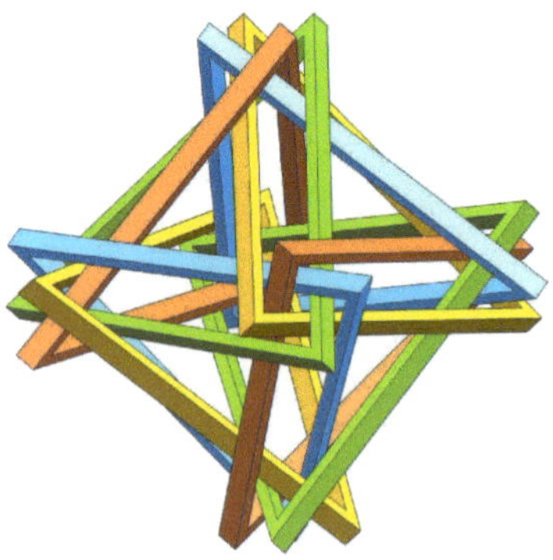

Download
https://grabcad.com/library/8-triangles-geometric-sculpture-1

1 edge of the triangle. x 24

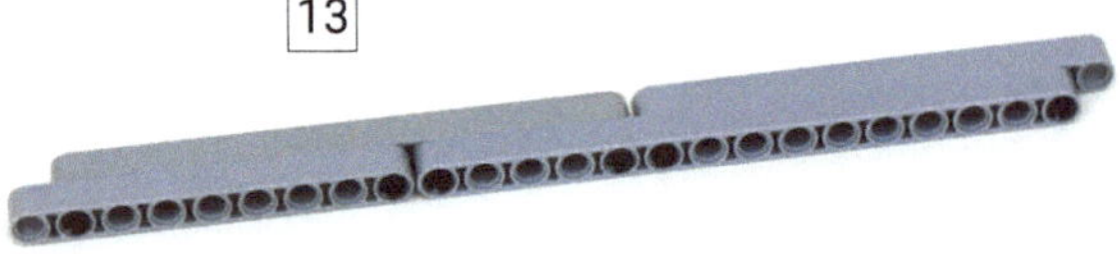

Parts:

15
x 24

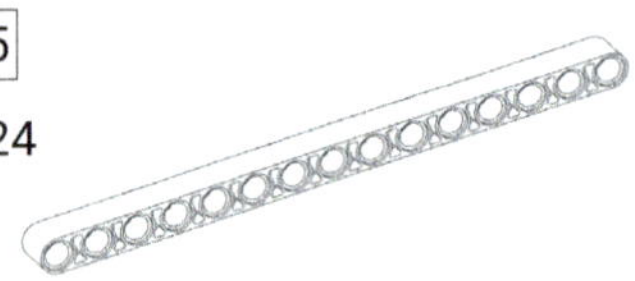

13
x 24

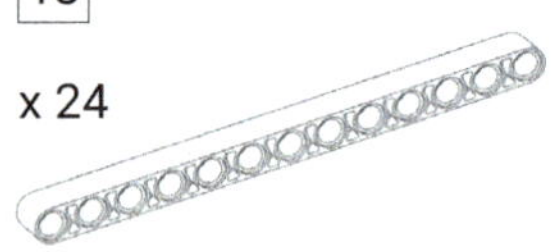

11

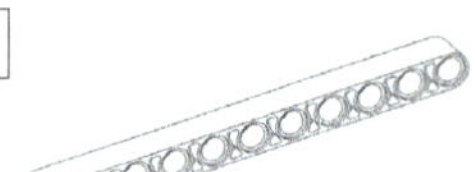

x 24

9

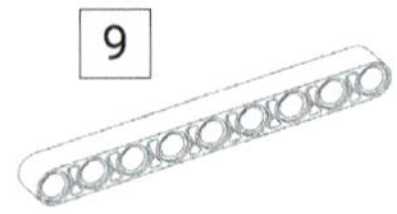

x 24

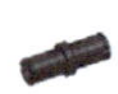

x 168

8 intersecting triangles.
Each triangle links with 3 others.

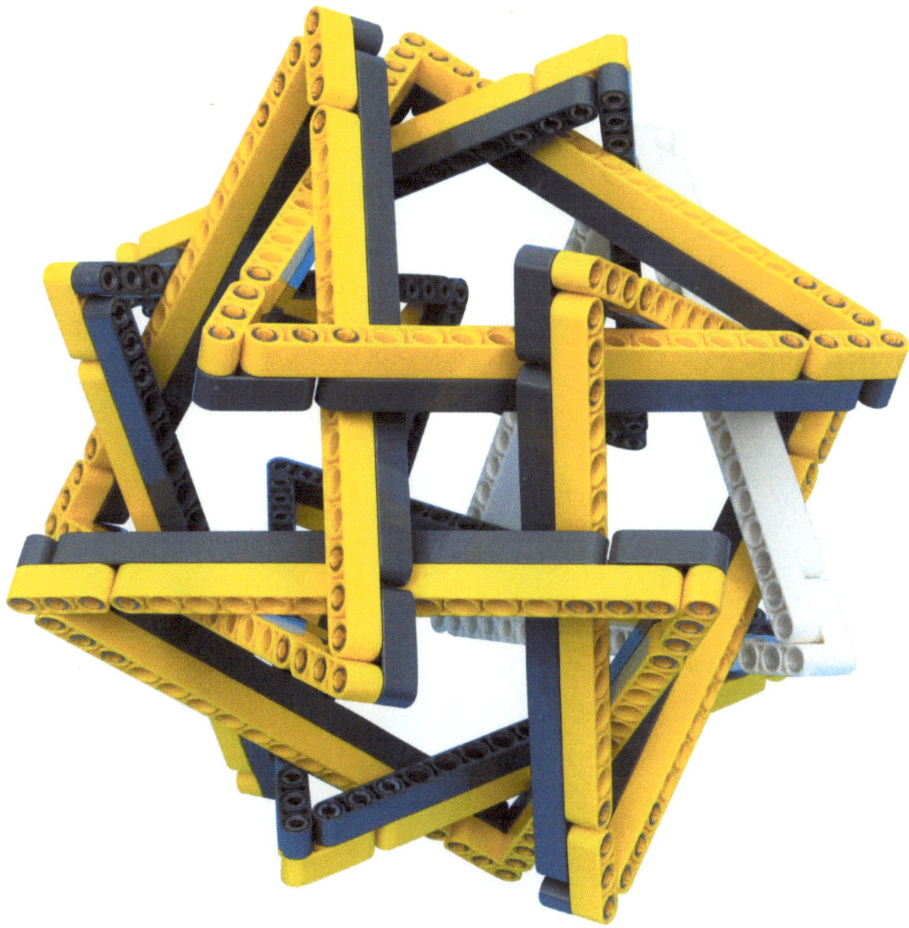

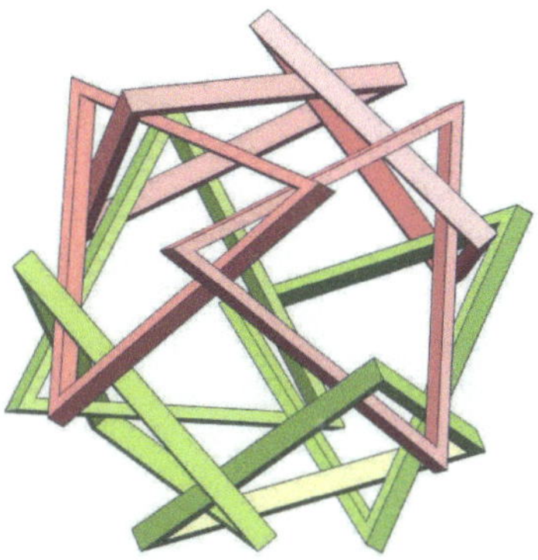

Download
https://grabcad.com/library/8-triangles-v2-geometric-sculpture-1

---

1 edge of the triangle.

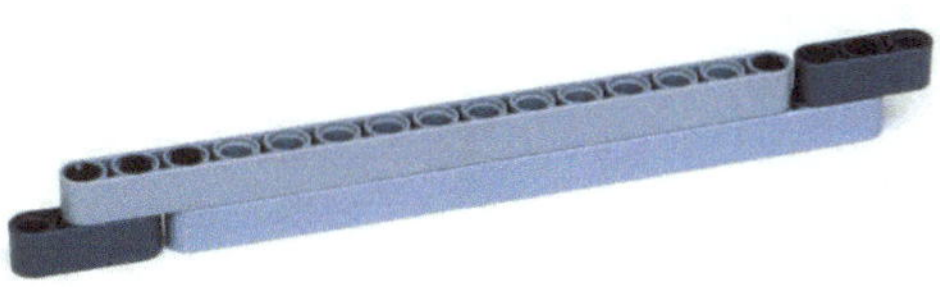

Parts:

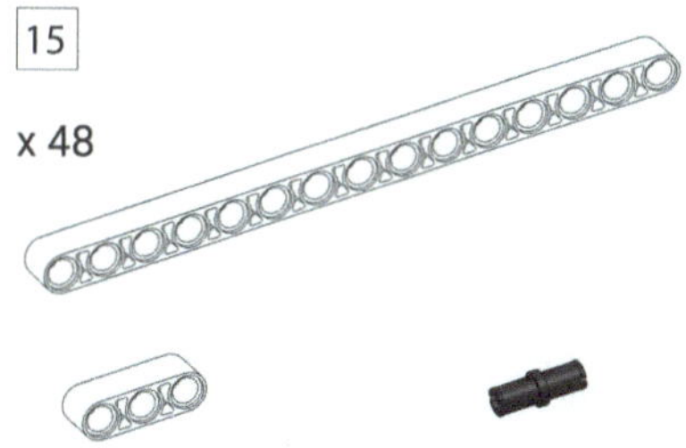

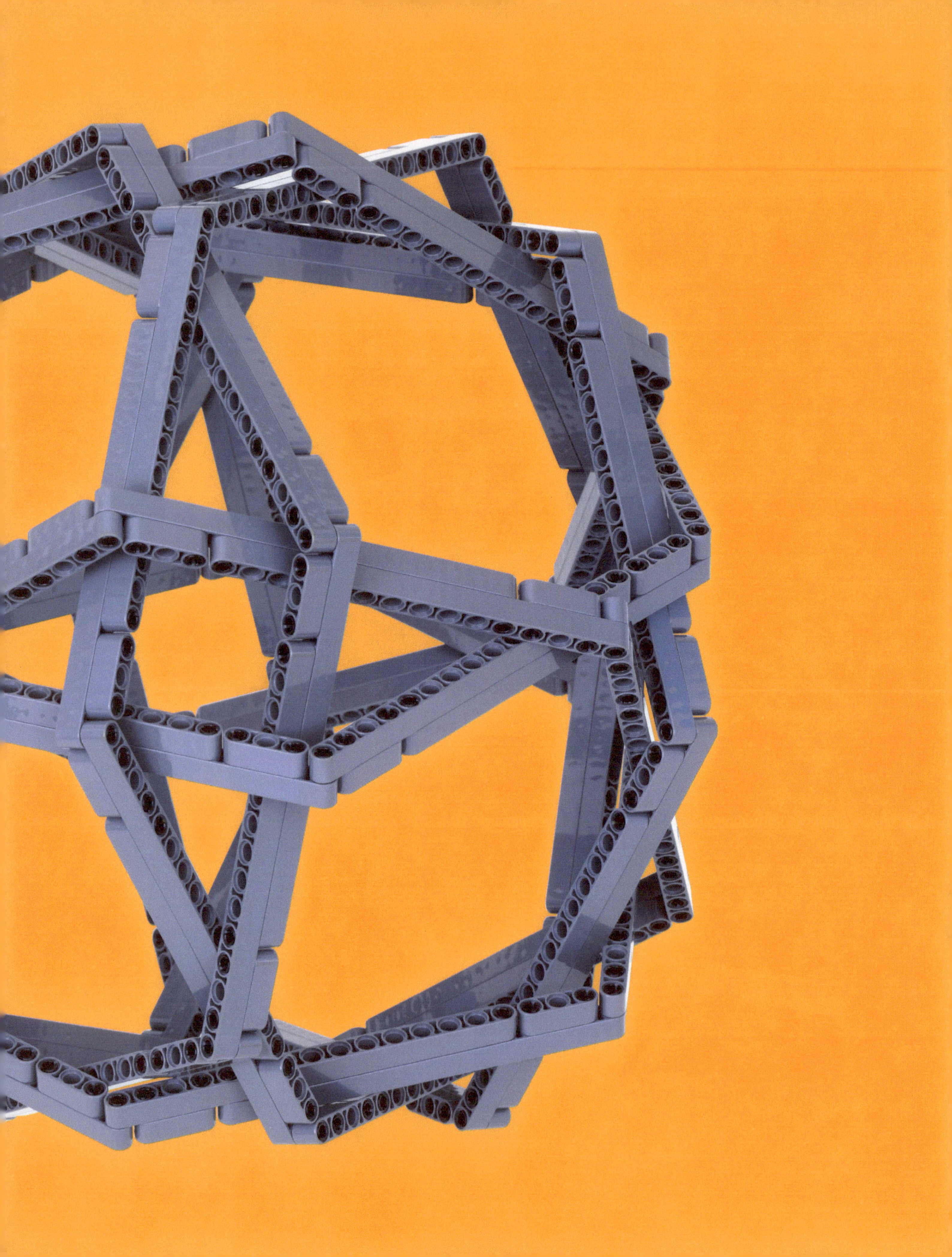

8 intersecting triangles.
This is based on the octahedron.

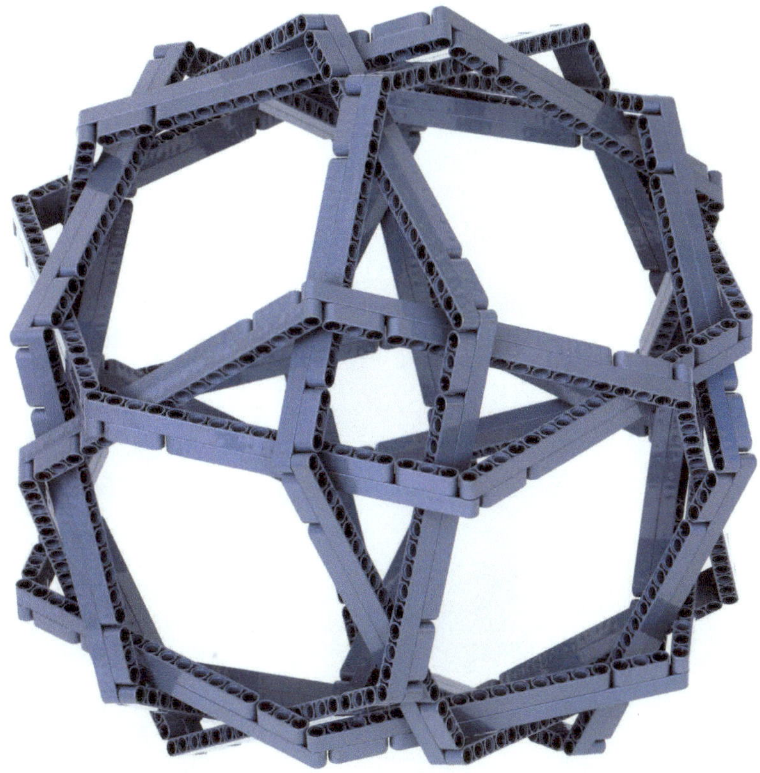

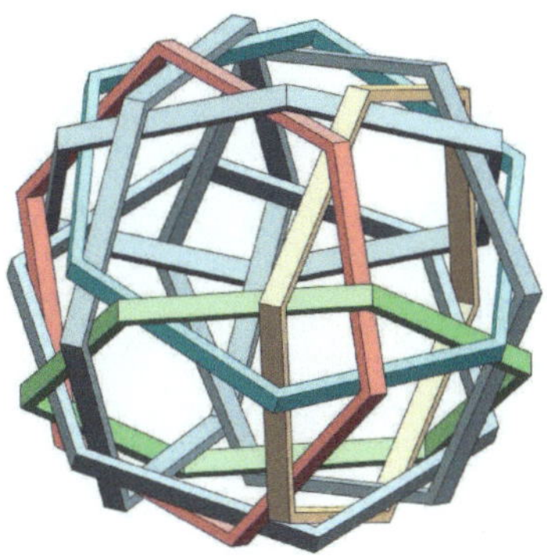

Download
https://grabcad.com/library/8-hexagons-geometric-sculpture-1

Parts:

15

x 96

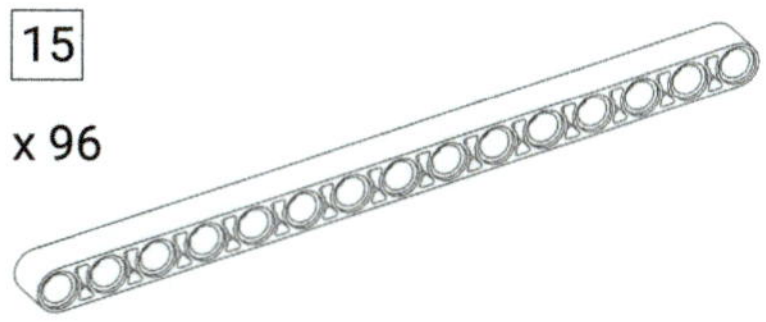

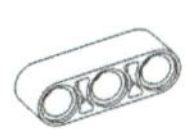

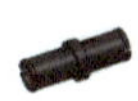

x 96

x 336

DODECAHEDRON

This section is about the dodecahedron and their related sculptures. A regular dodecahedron is composed of twelve regular pentagonal faces, three meeting at each vertex. It is one of the five Platonic solids. It has 12 faces, 20 vertices and 30 edges.

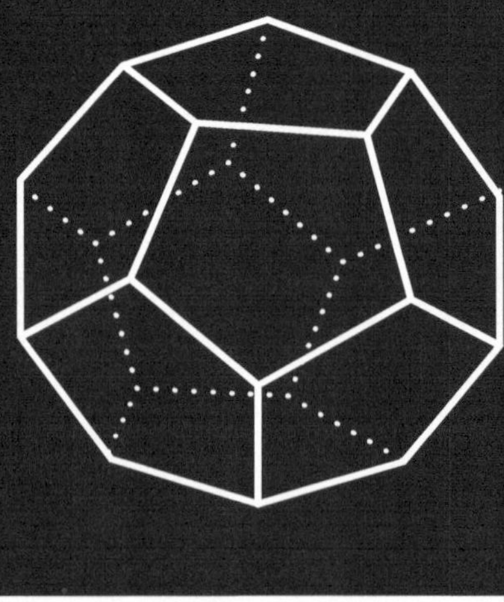

CHAPTER 4

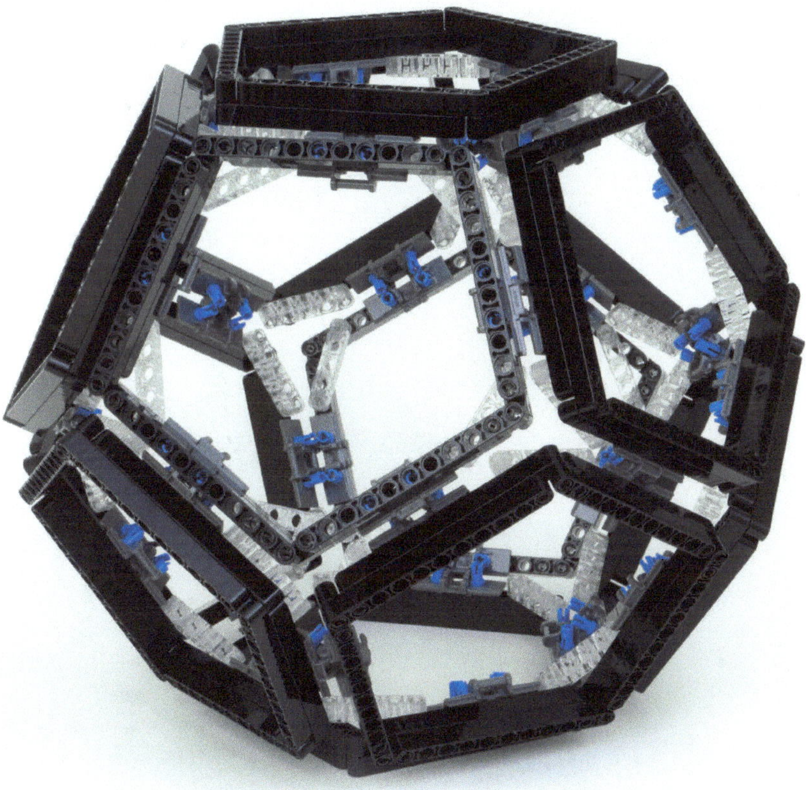

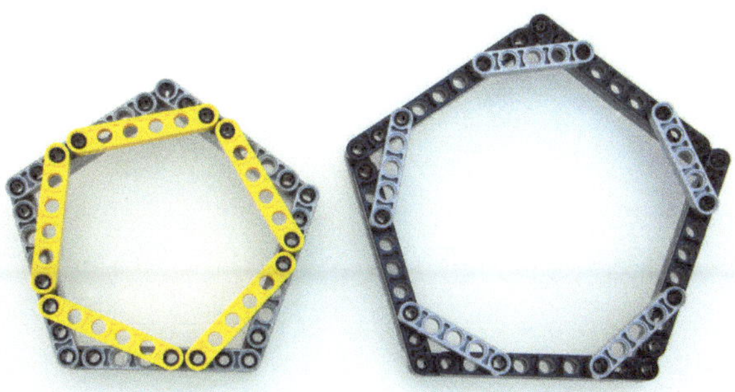

A study on 2 types of pentagon construction - The inside angle of a pentagon is 108°. The one on the left is formed by triangles of ratio 2-4-5 at each corner and has an angle of 108.2° . The right side pentagon corner triangles are of ratio 2-3-4 and has an angle of 104.5°and it still fits in. This is used for the construction of the dodecahedron.

x 12

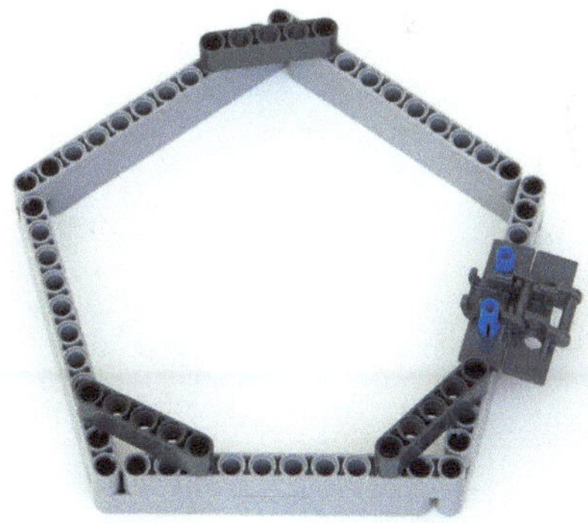

Parts:

11 x 120

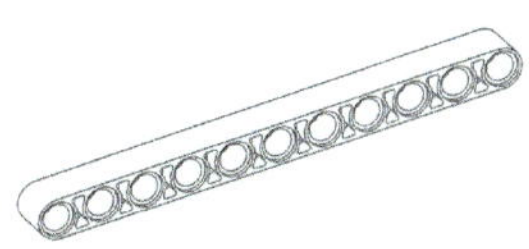

5 x 36

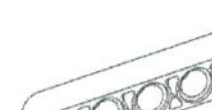

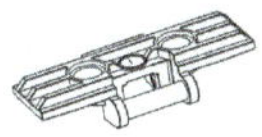

x 60

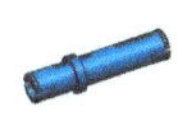

x 120

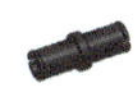

x 252

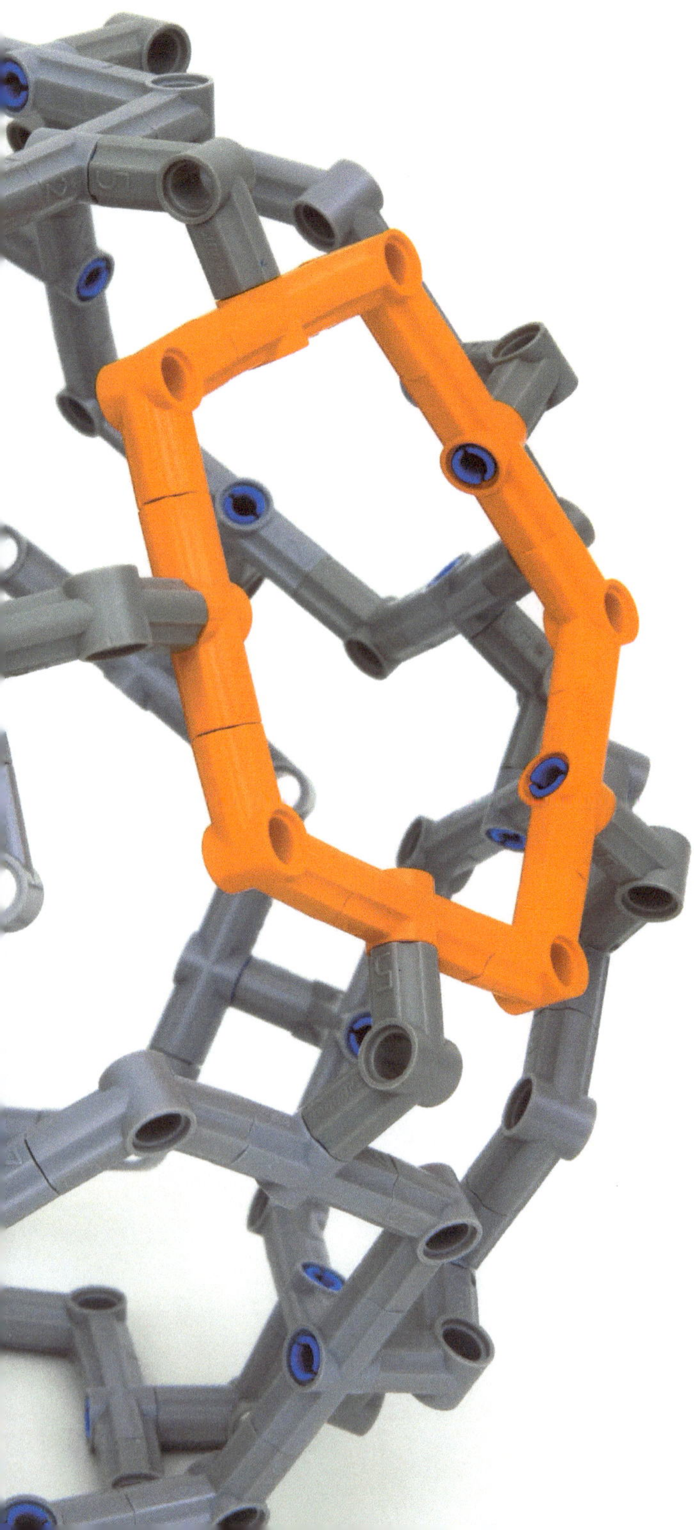

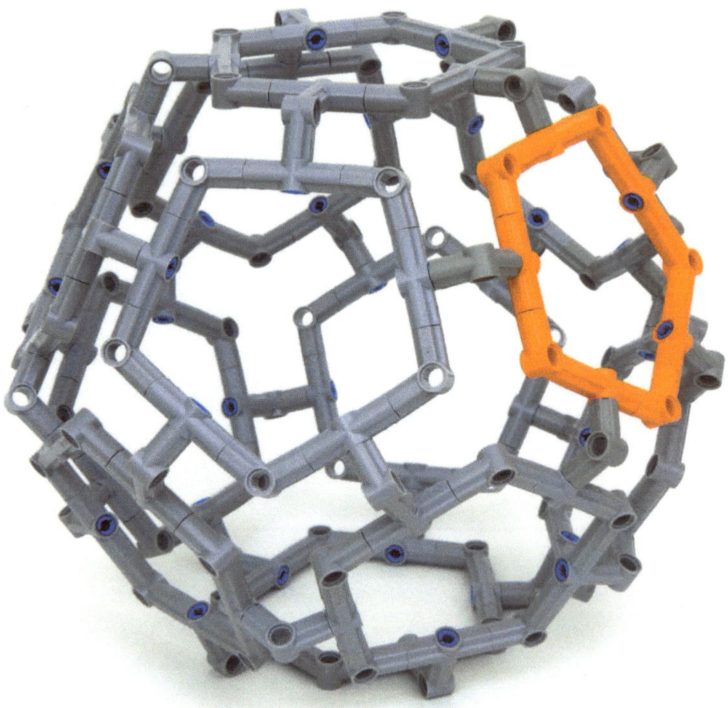

Parts:

#5

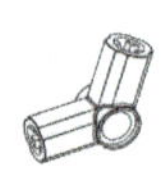

x 90

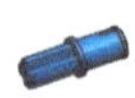

x 60

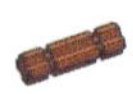

x 120

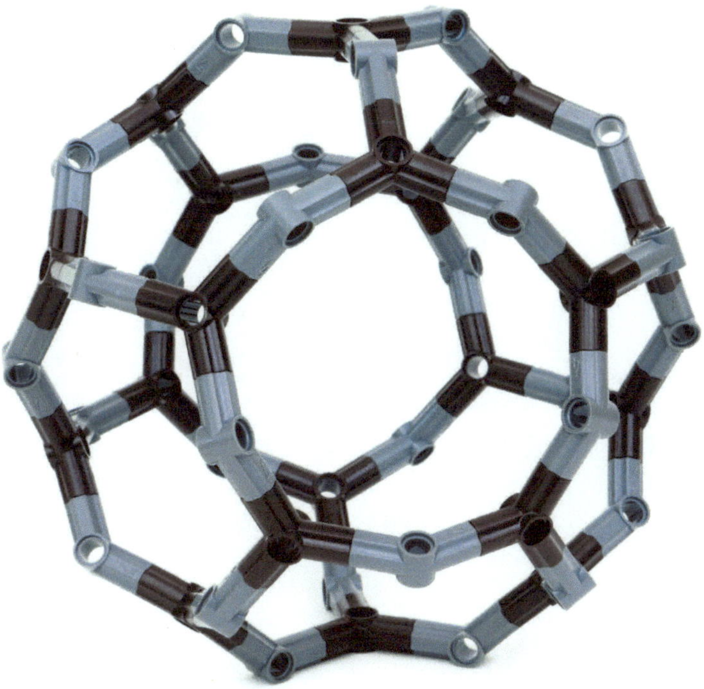

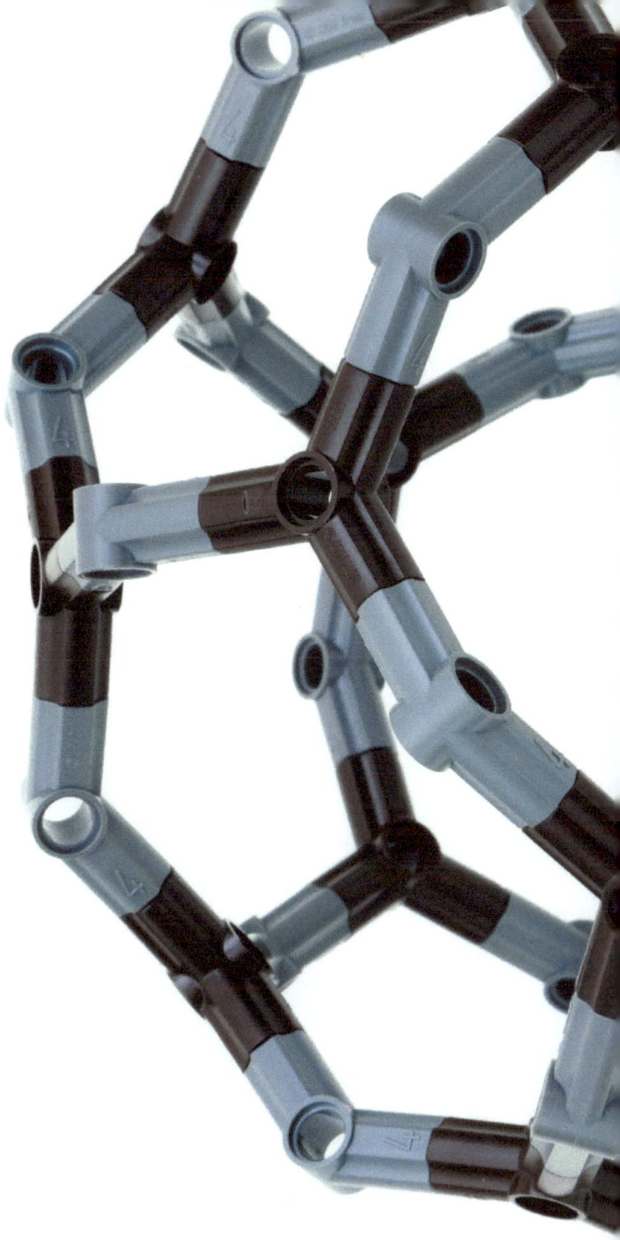

Parts:

#4

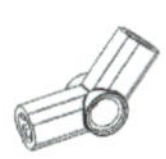

x 30

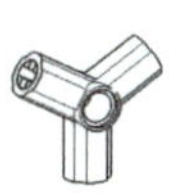

x 20

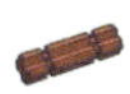

x 60

This sculpture is derived from the 5 intersecting rectangles that spans the opposite faces of a dodecahedron.

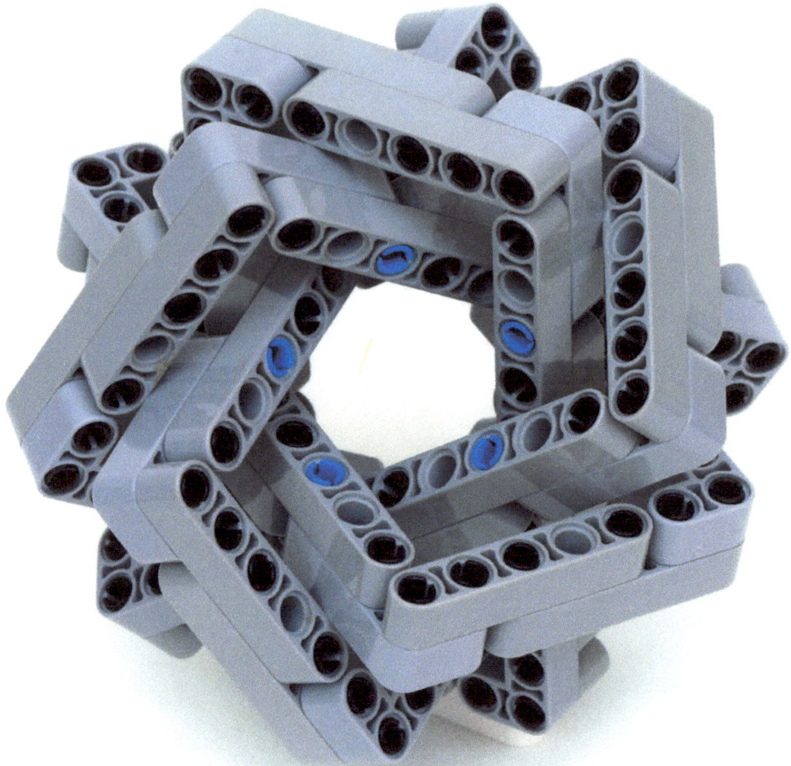

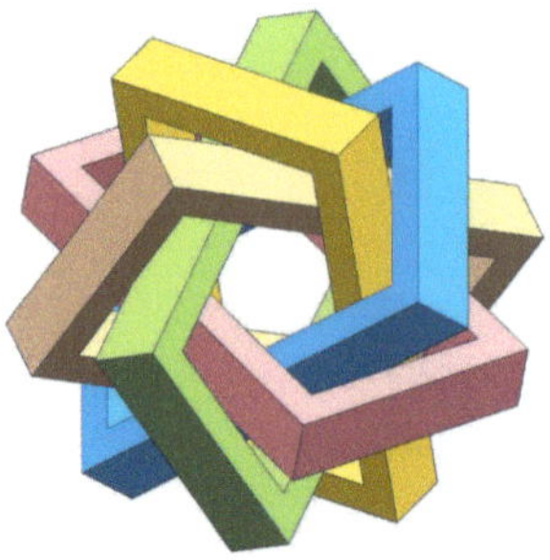

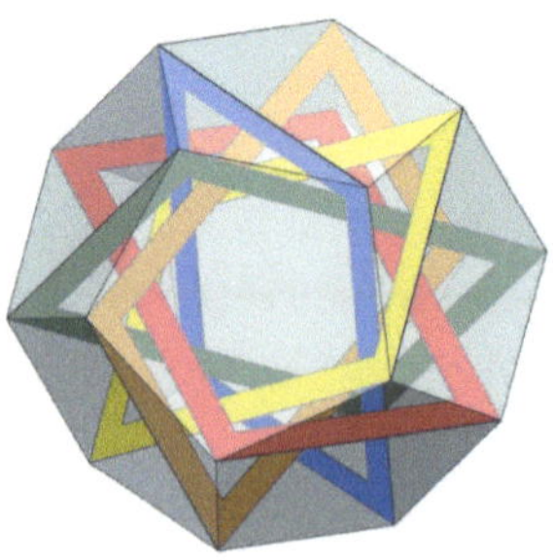

Download

https://grabcad.com/library/5-rectangles-geometric-sculpture-1

x 5

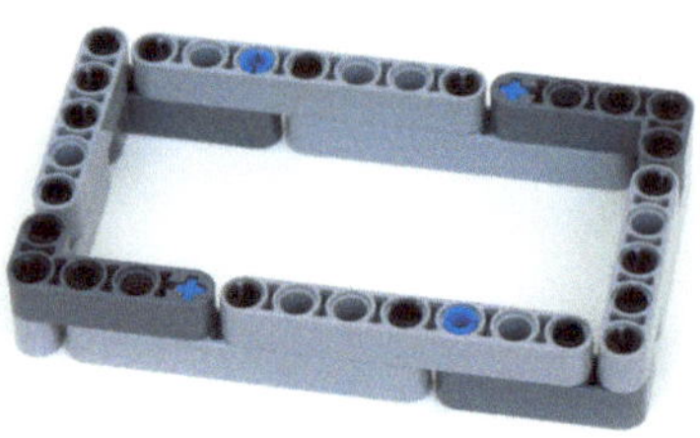

Parts:

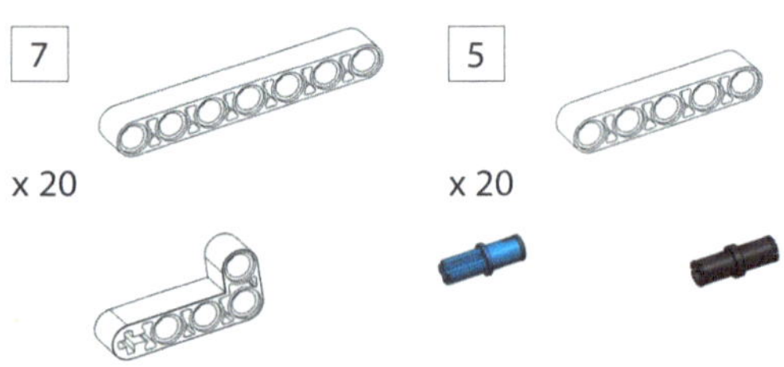

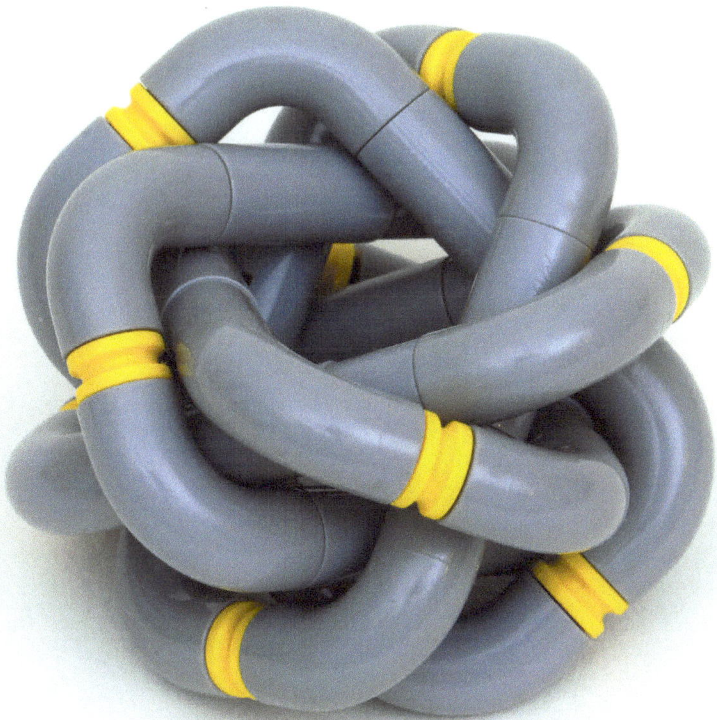

5 Intersecting rectangles - a smaller version.

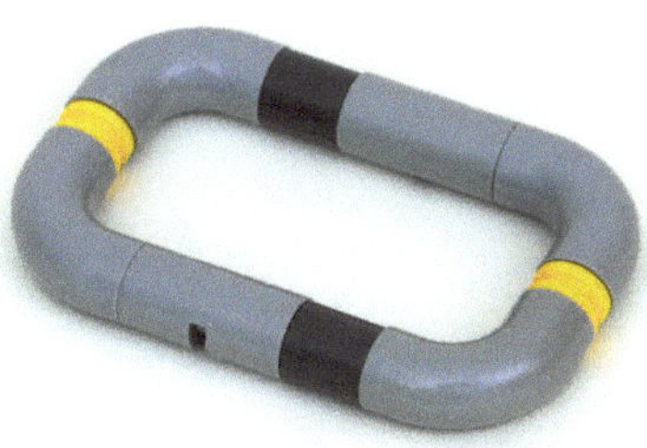

Parts:

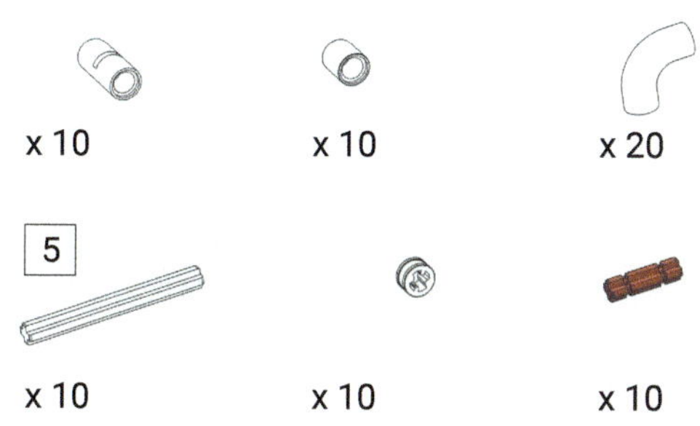

5 intersecting tetrahedra also known as compound of 5 tetrahedra.
This is a well known geometric construction.

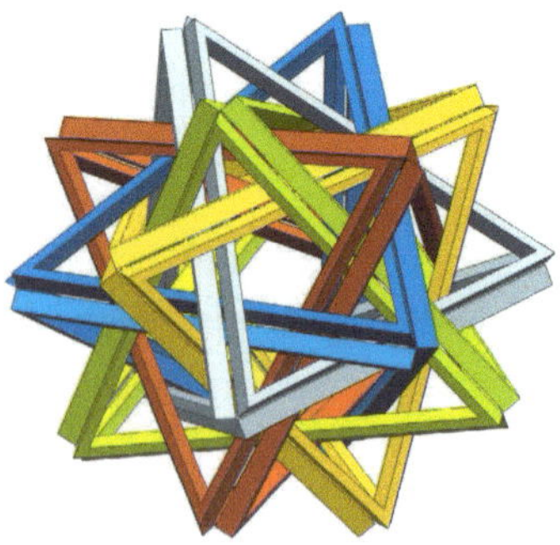

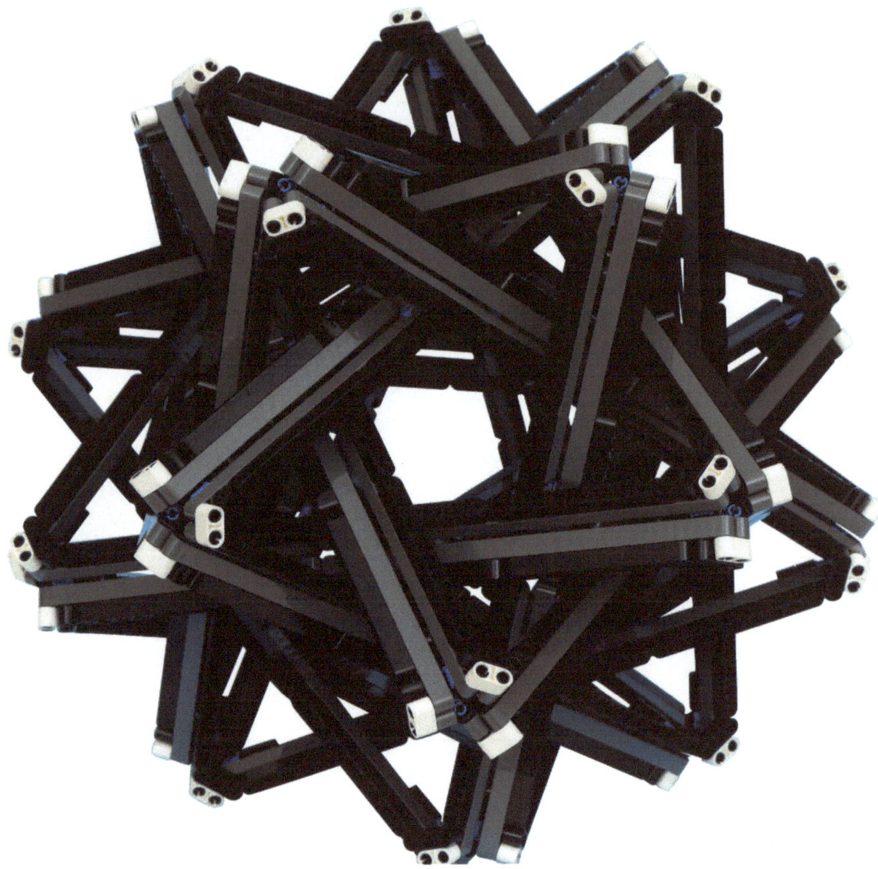

Download
https://grabcad.com/library/five-intersecting-tetrahedra-1

1 edge of the tetrahedron - total 30 edges for the sculpture.
x 30

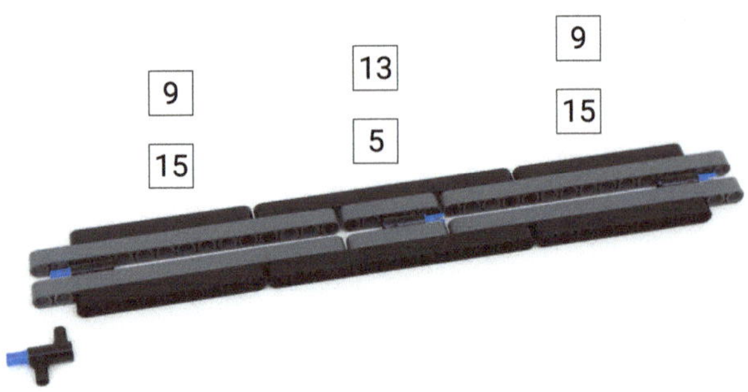

Parts:

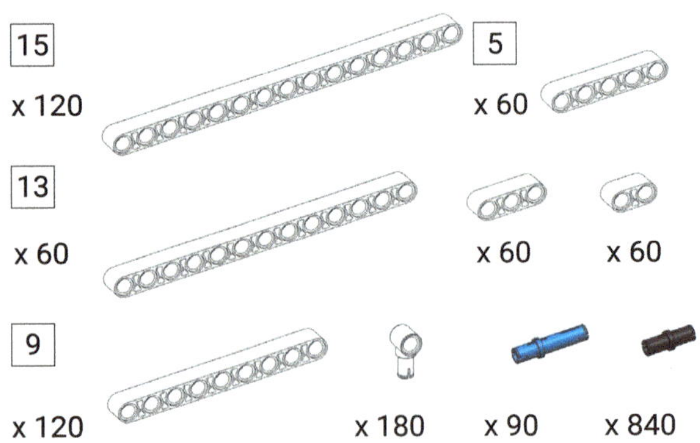

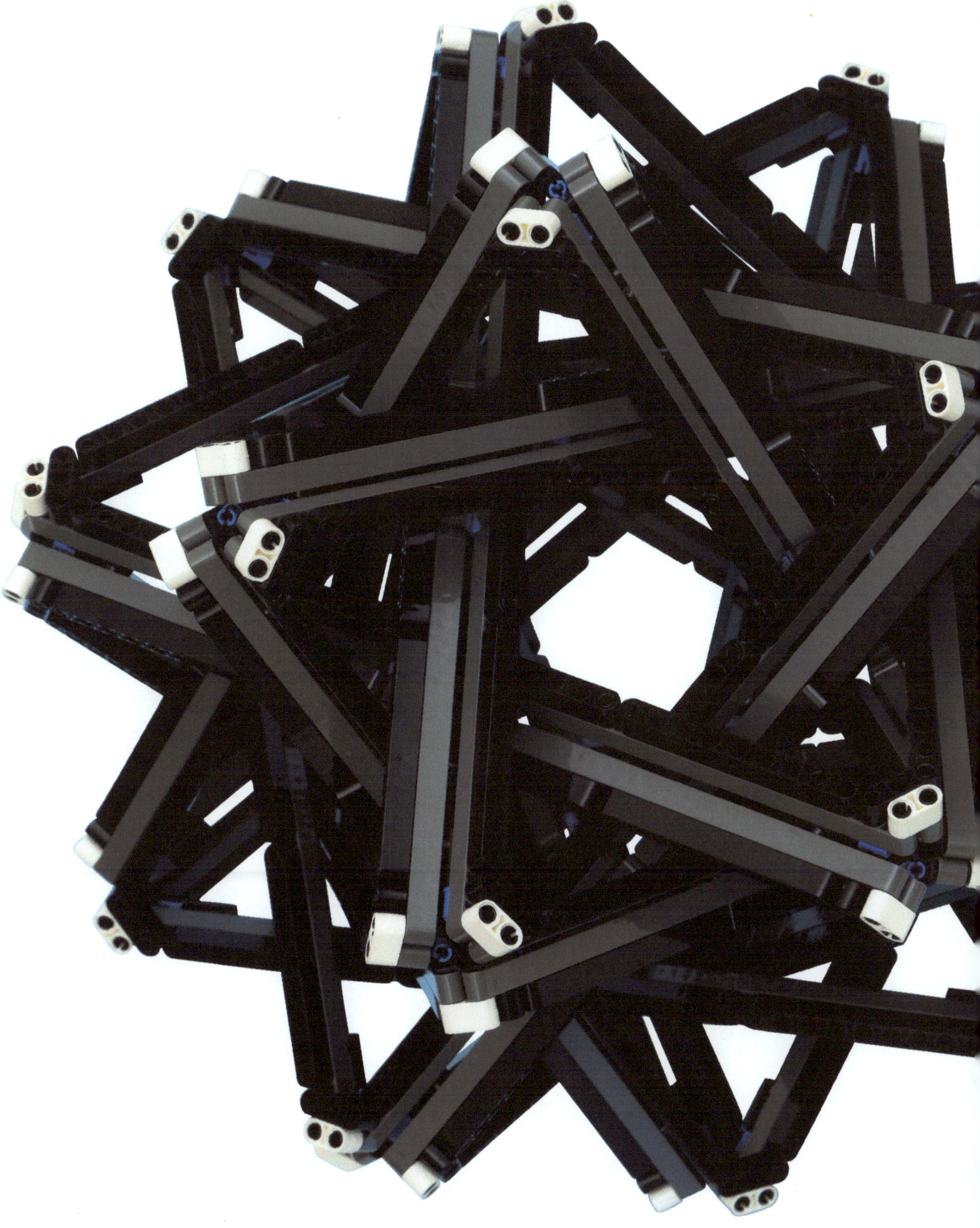

6 intersecting pentagons – This sculpture is derived from the dodecahedron where the opposing faces are simultaneously rotated and translated to meet at the center.

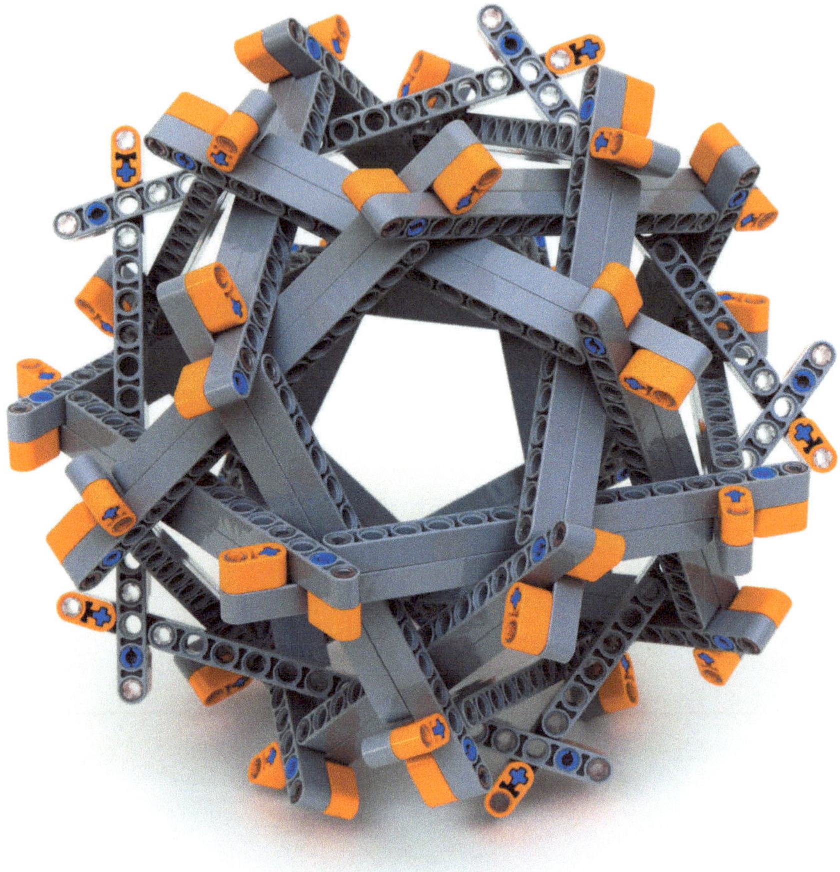

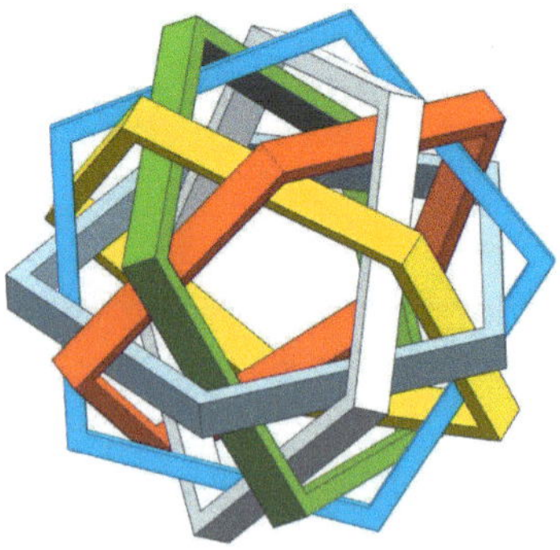

Download
https://grabcad.com/library/6-intersecting-pentagons-geometric-sculpture-1

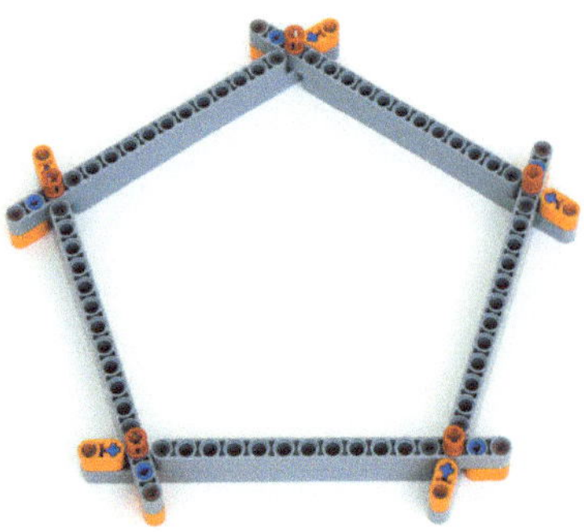

Parts:

15

x 60

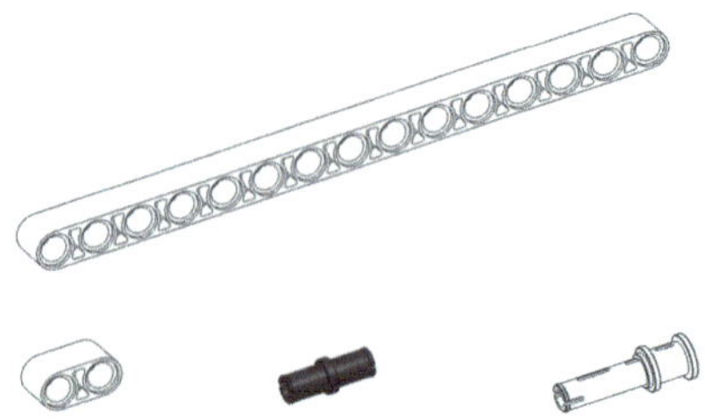

x 60 x 180 x 20

12 intersecting pentagons. This sculpture is derived from the dodecahedron where the opposing faces are simultaneously rotated and translated to intersect at the edges.

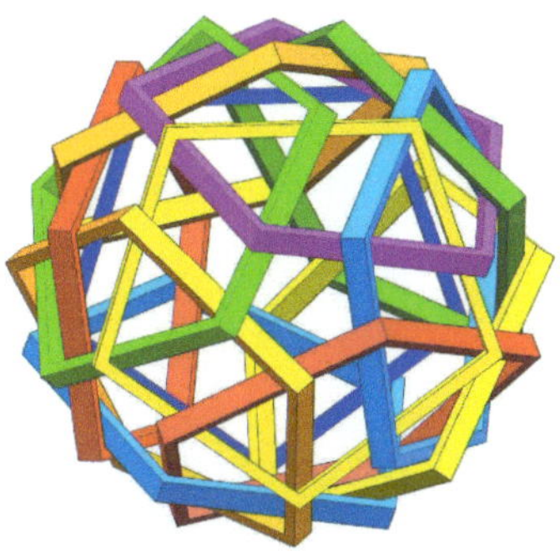

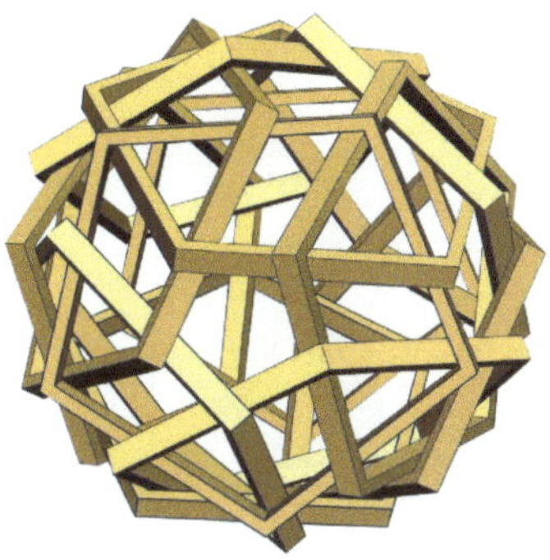

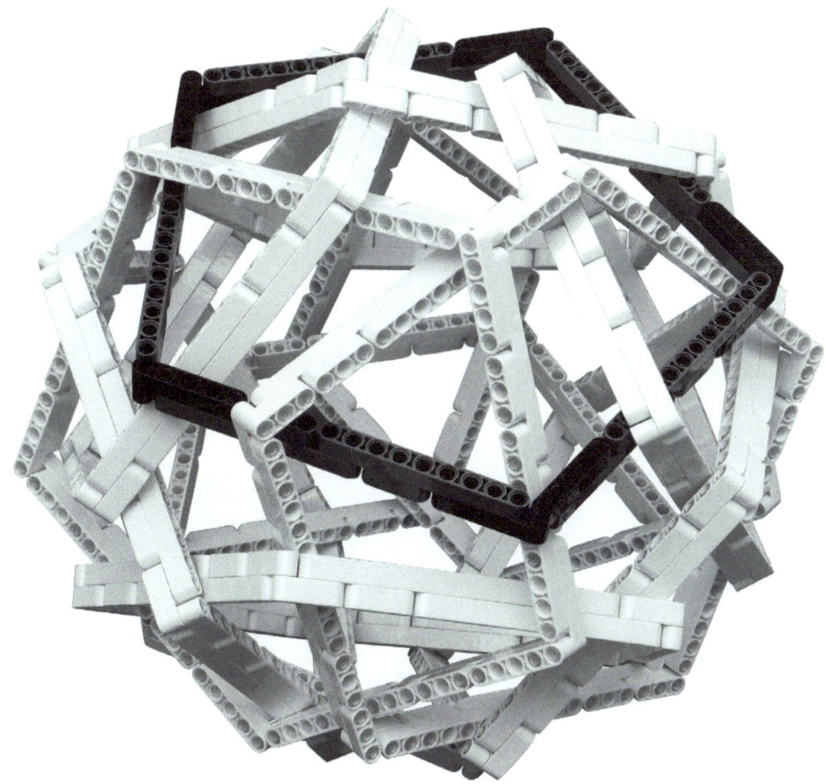

Download
https://grabcad.com/library/12-pentagons-geometric-sculpture-1

1 edge of the pentagon.

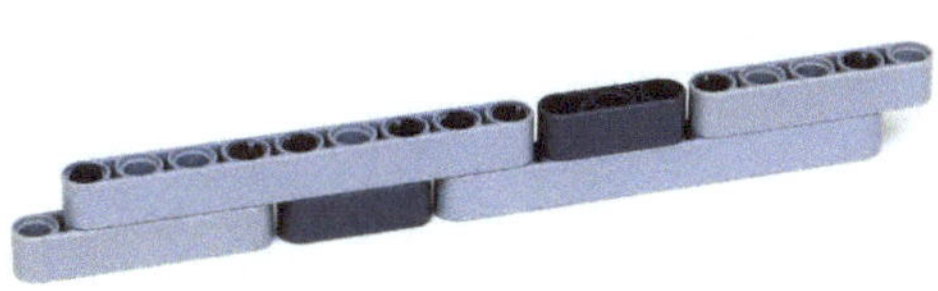

Parts:

x 120 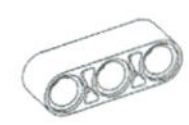

9

x 120 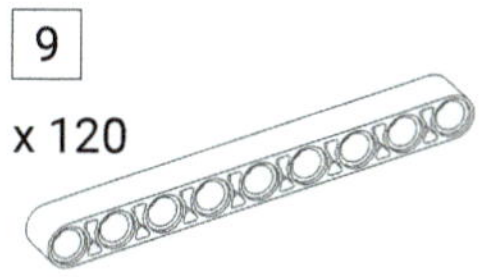

5

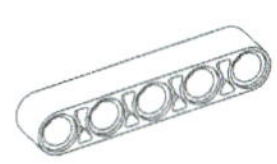

x 120

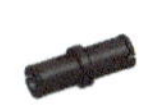

x 660

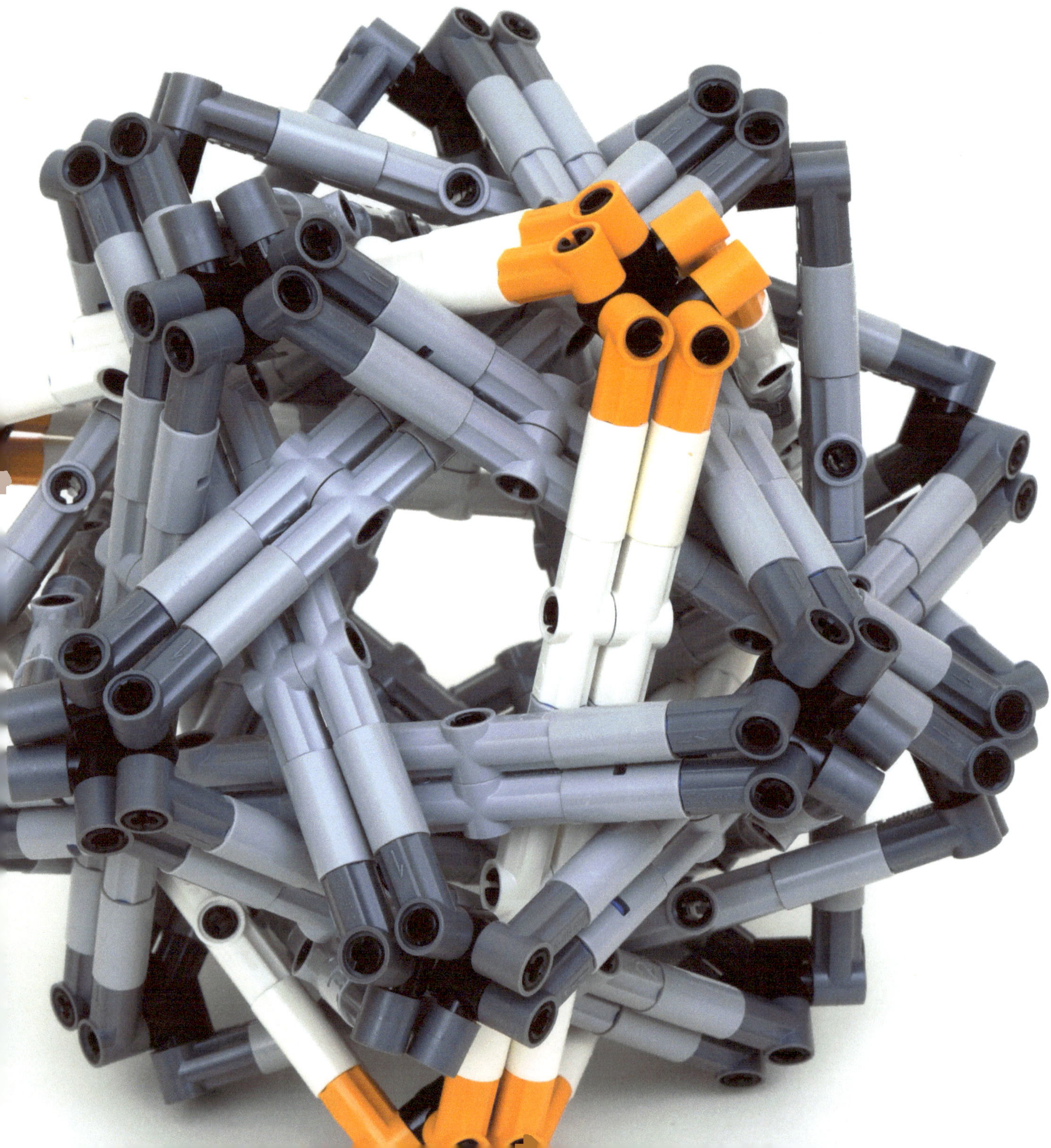

The 5 tetrahedra are constructed inside the dodecahedron by connecting the vertices. There are also other ways to construct 5 intersecting tetrahedra.

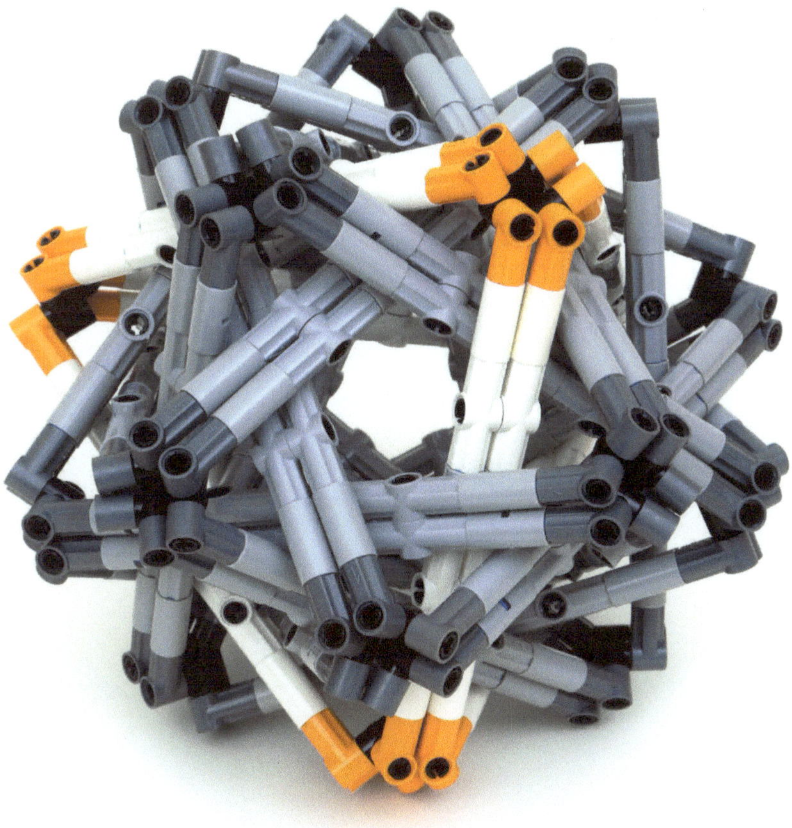

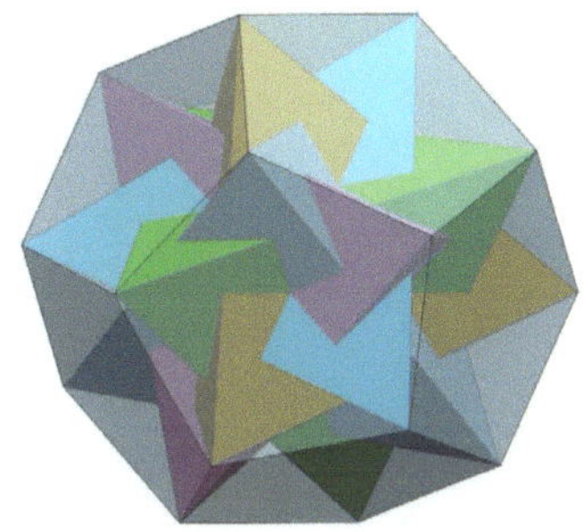

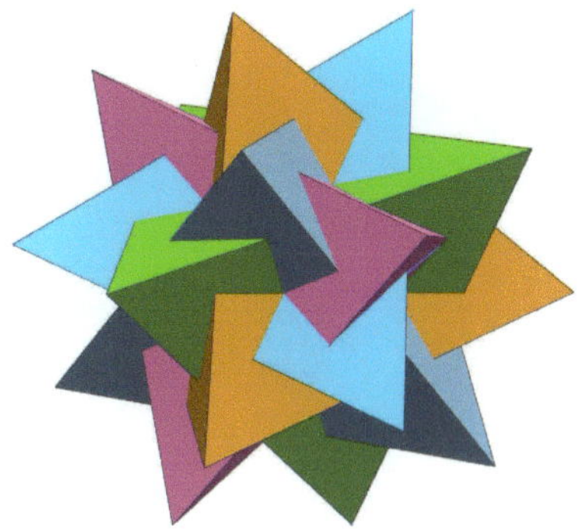

A smaller 5 intersecting tetrahedra.

1 edge of the tetrahedron. x 30

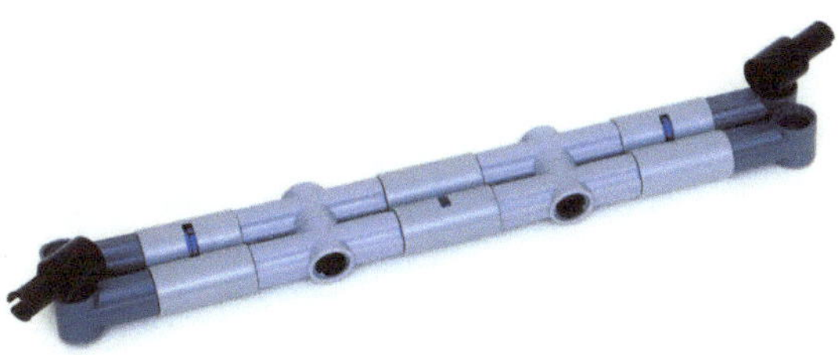

Parts:

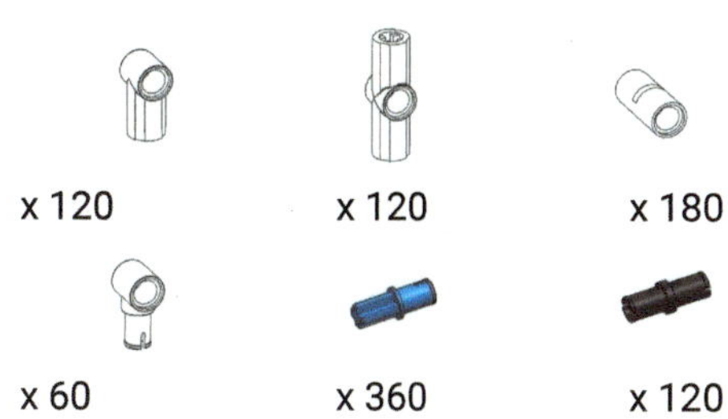

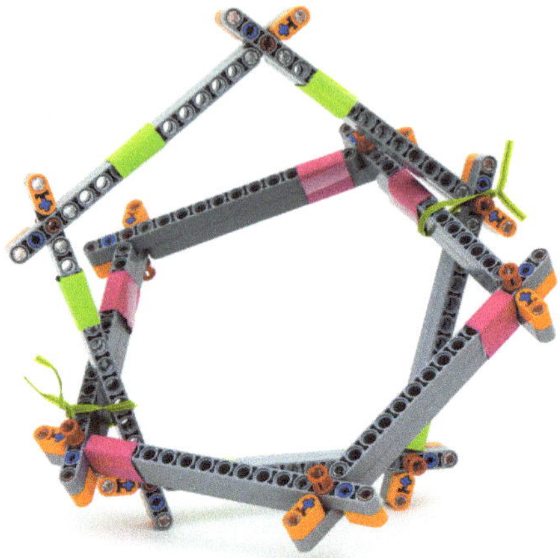

continue from page 62

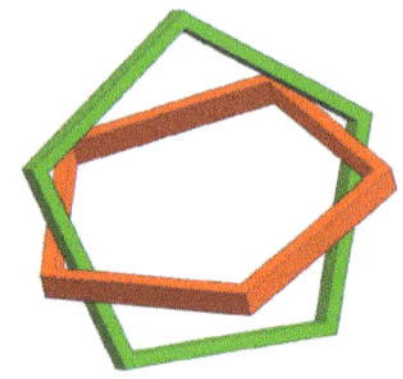

Remove the wires and holding pins after completion. This is an interlocking structure and check that the adjacent beams are in the correct orientation.

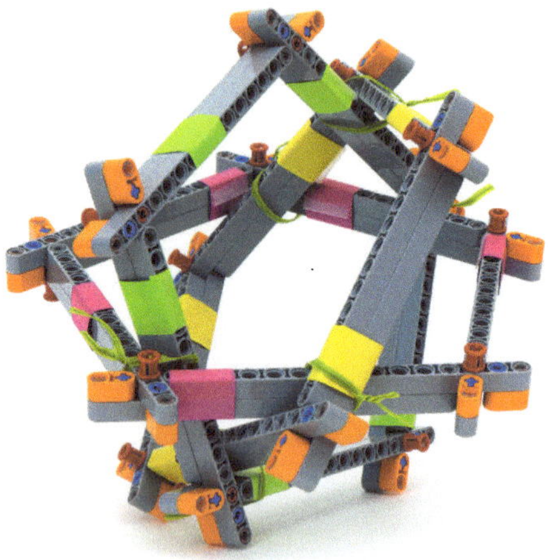

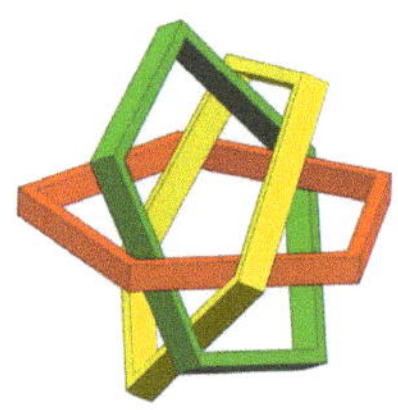

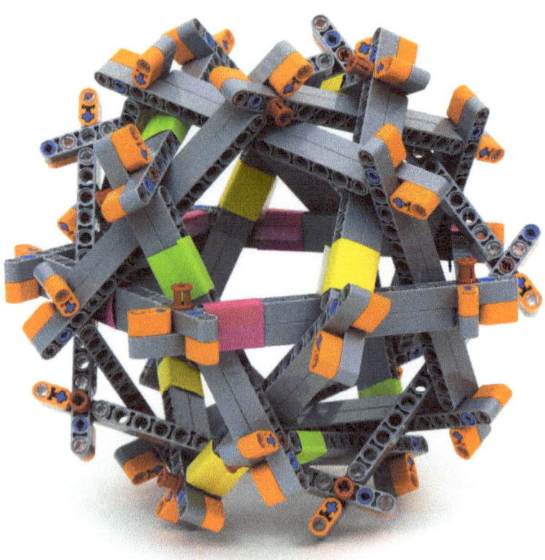

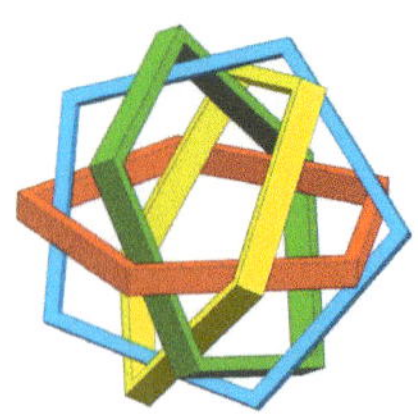

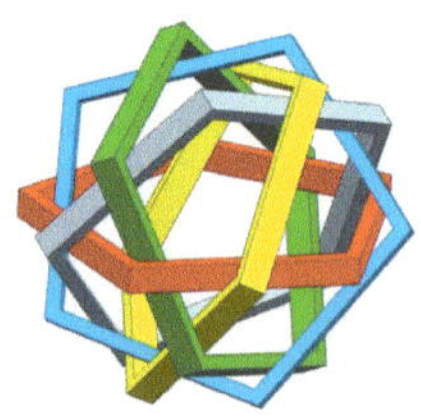

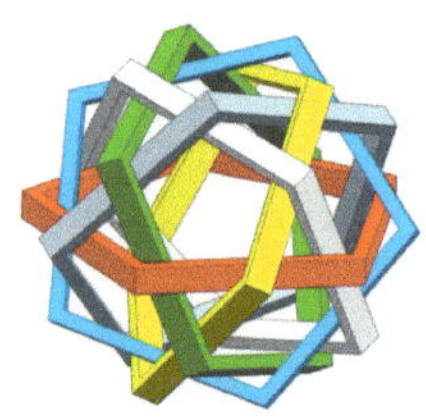

This version uses 60 nos. 1 x 13 beams.

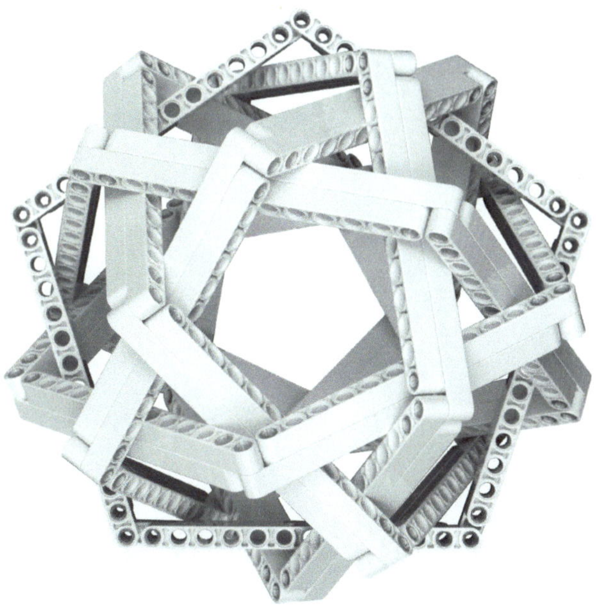

6 intersecting pentagons using connectors.

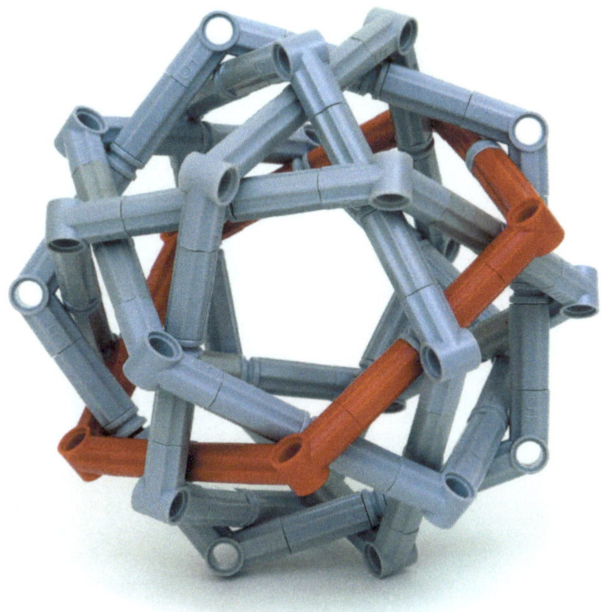

Assembly of 1 pentagon. The angled connector is 112.5° and the internal angle of the pentagon is 108° . There is a small error but the assembly is flexible to fit. Spaced the red pin equally on the half spacer.

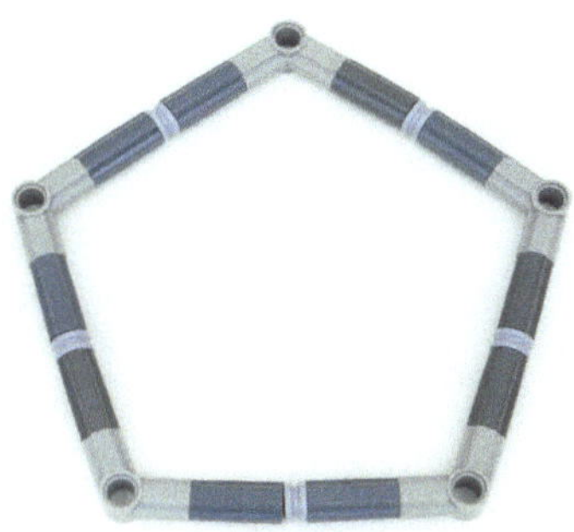

Parts:

5

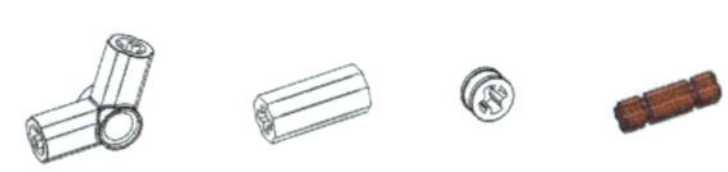

x 30 x 60 x 30 x 90

CHAPTER 5

ICOSAHEDRON

A regular icosahedron is a polyhedron with 20 faces, 30 edges and 12 vertices. It is one of the five Platonic solids, and the one with the most edges.

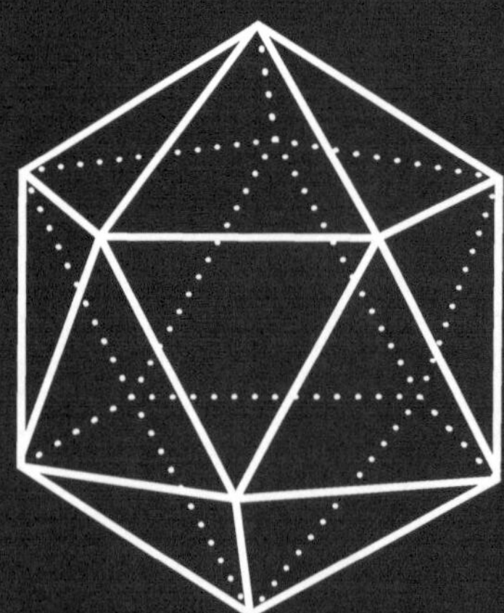

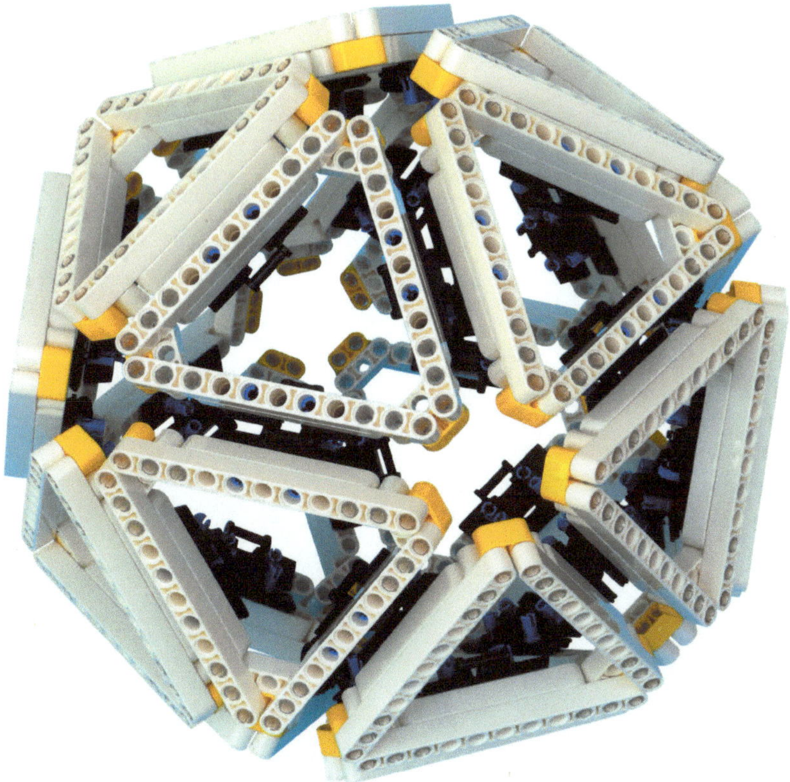

20 triangles in an icosahedron configuration. Each face is an equilateral triangle.

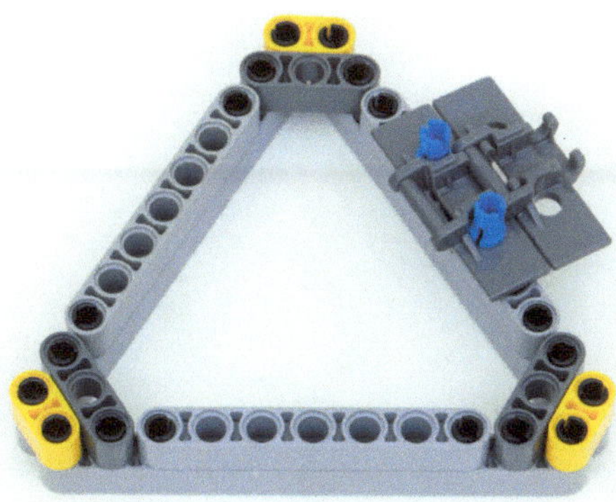

Assembly of 1 triangle and hinge.

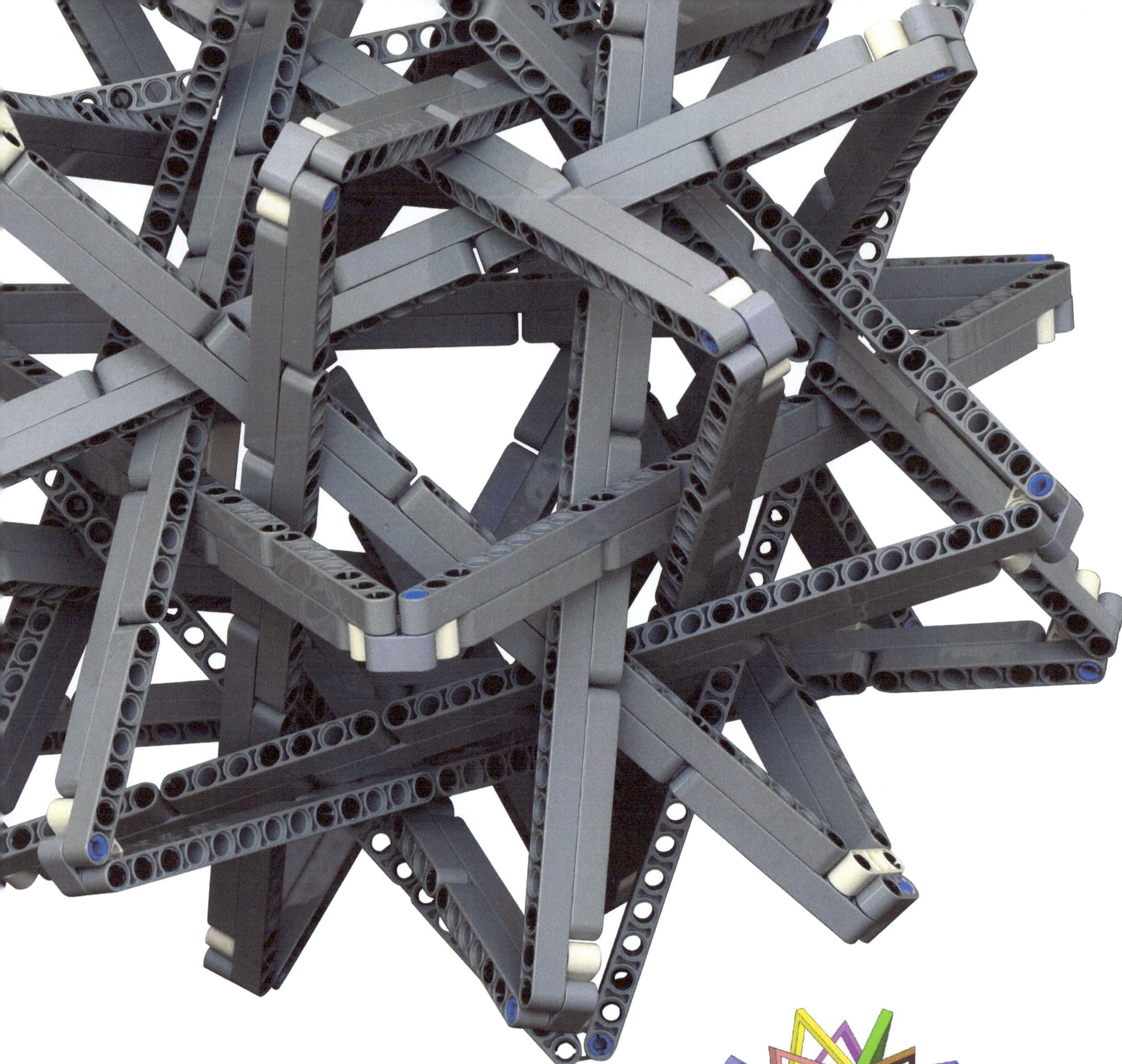

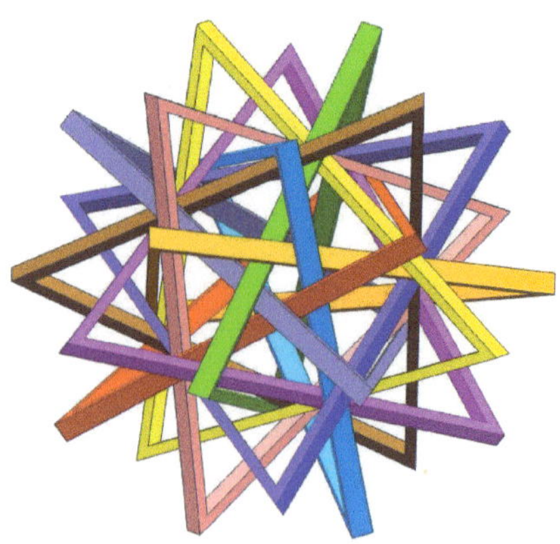

Download
https://grabcad.com/ library/ten-triangles-geometric-sculpture-1

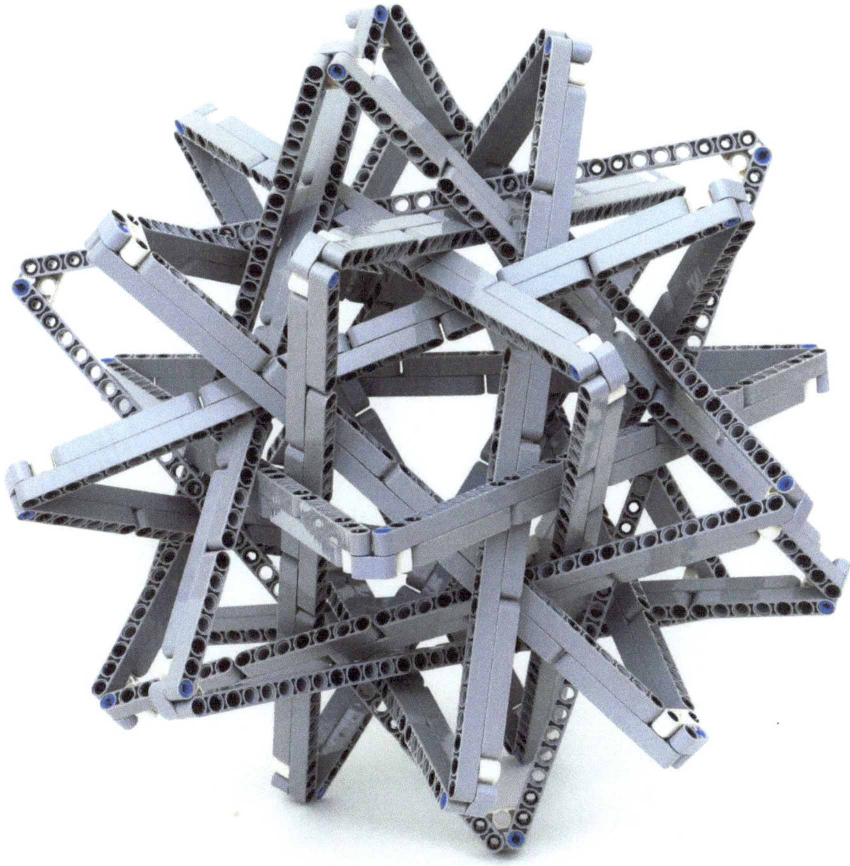

10 intersecting triangles. This is derived from the icosahedron by rotating and translating the opposing faces to align and meet at the center.

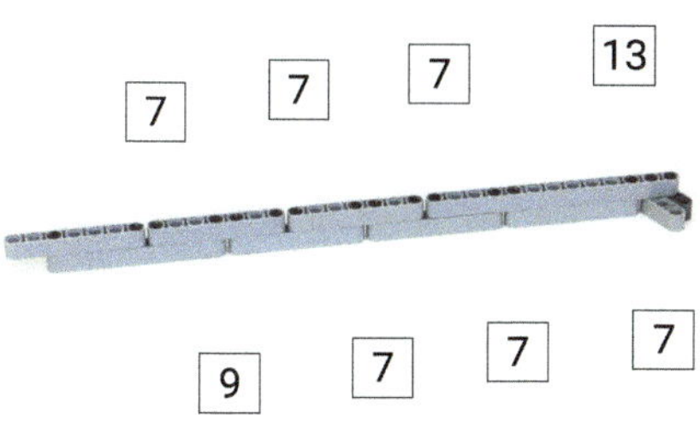

Parts:

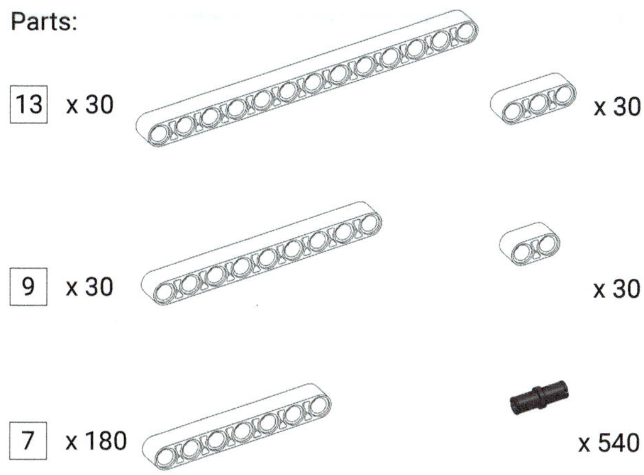

20 Intersecting triangles is made by simultaneously rotating the faces of the icosahedron.

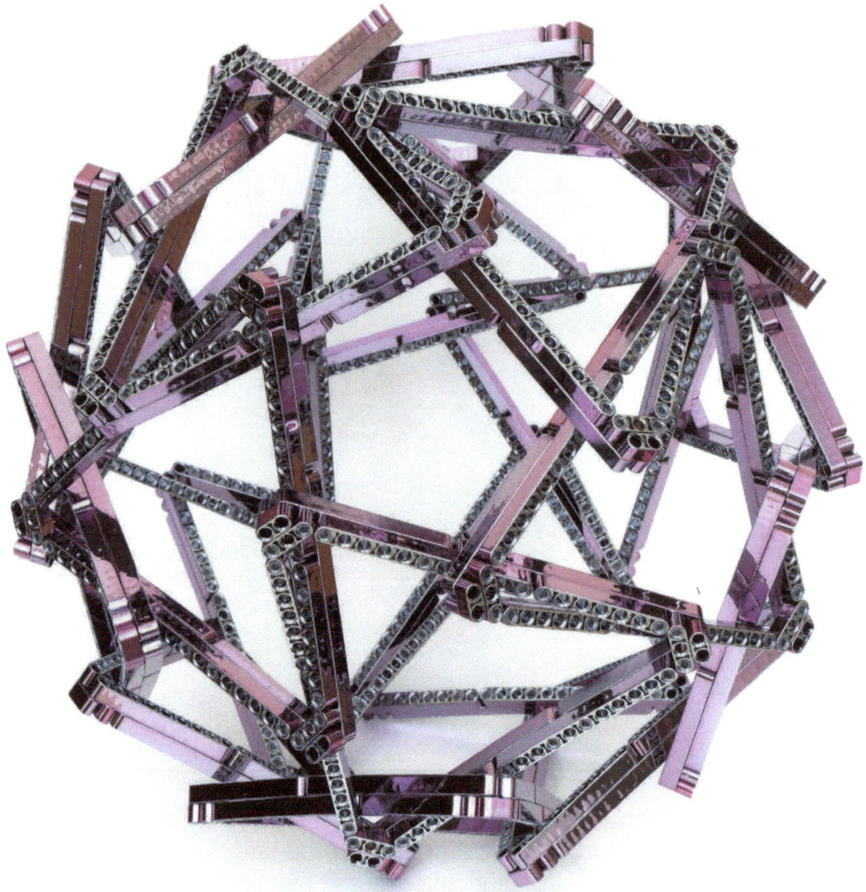

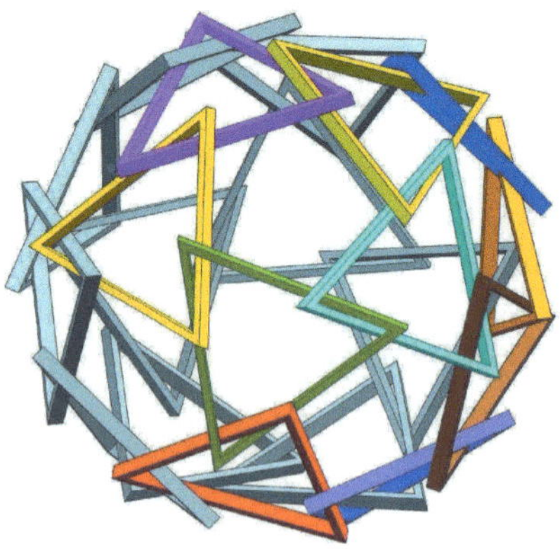

Download
https://grabcad.com/library/20-triangles-geometric-sculpture-1

1 triangle edge.

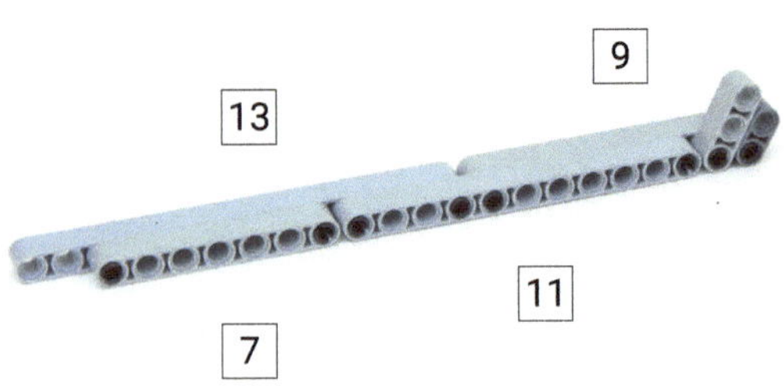

Parts:

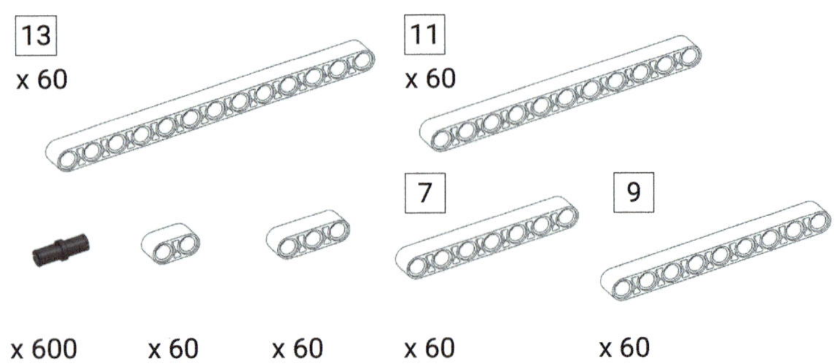

A Truncated Icosahedron formed by 10 nonagons.

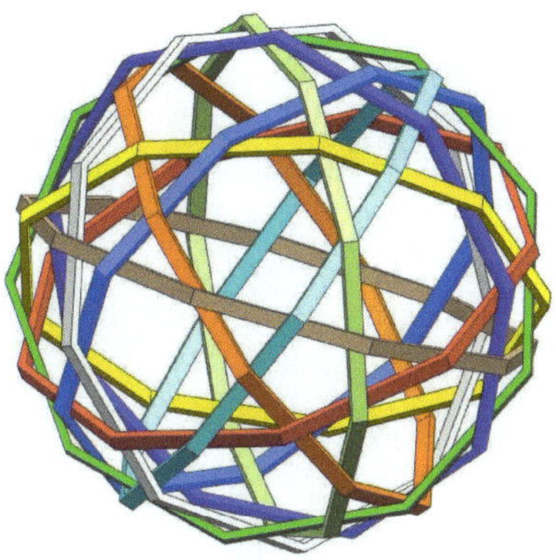

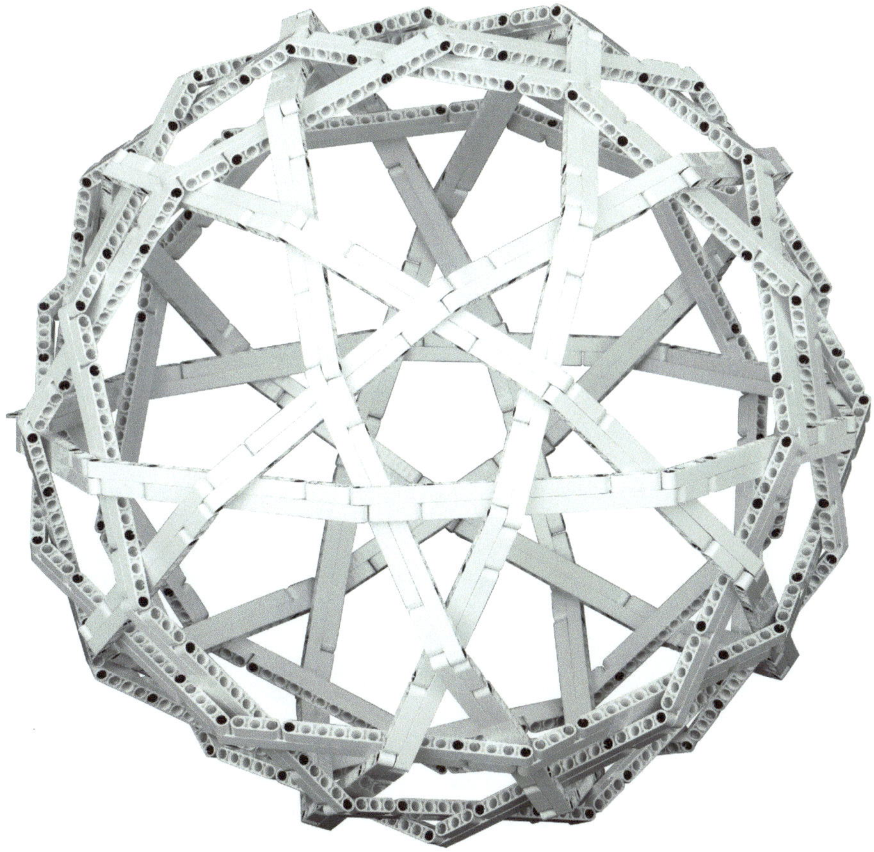

Download
https://grabcad.com/library/truncated-icosahedron-geometric-sculpture-1

1 edge of the nonagon. 90 x required.

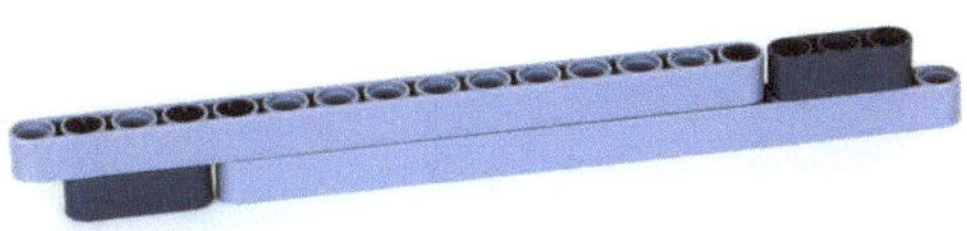

Parts:

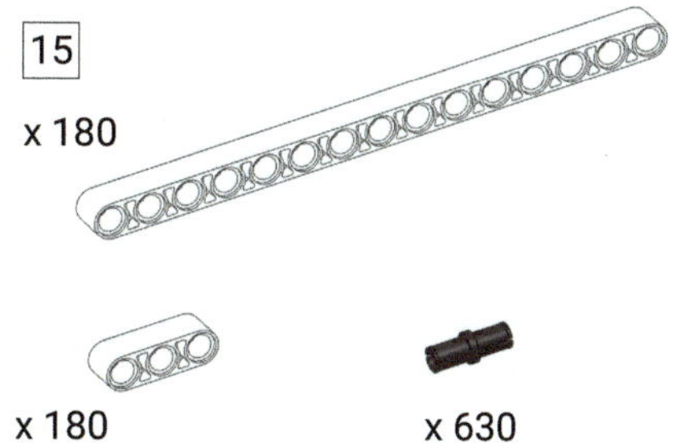

A Truncated Icosahedron also known as Buckyball, Fullerene C60 or Soccer Ball.

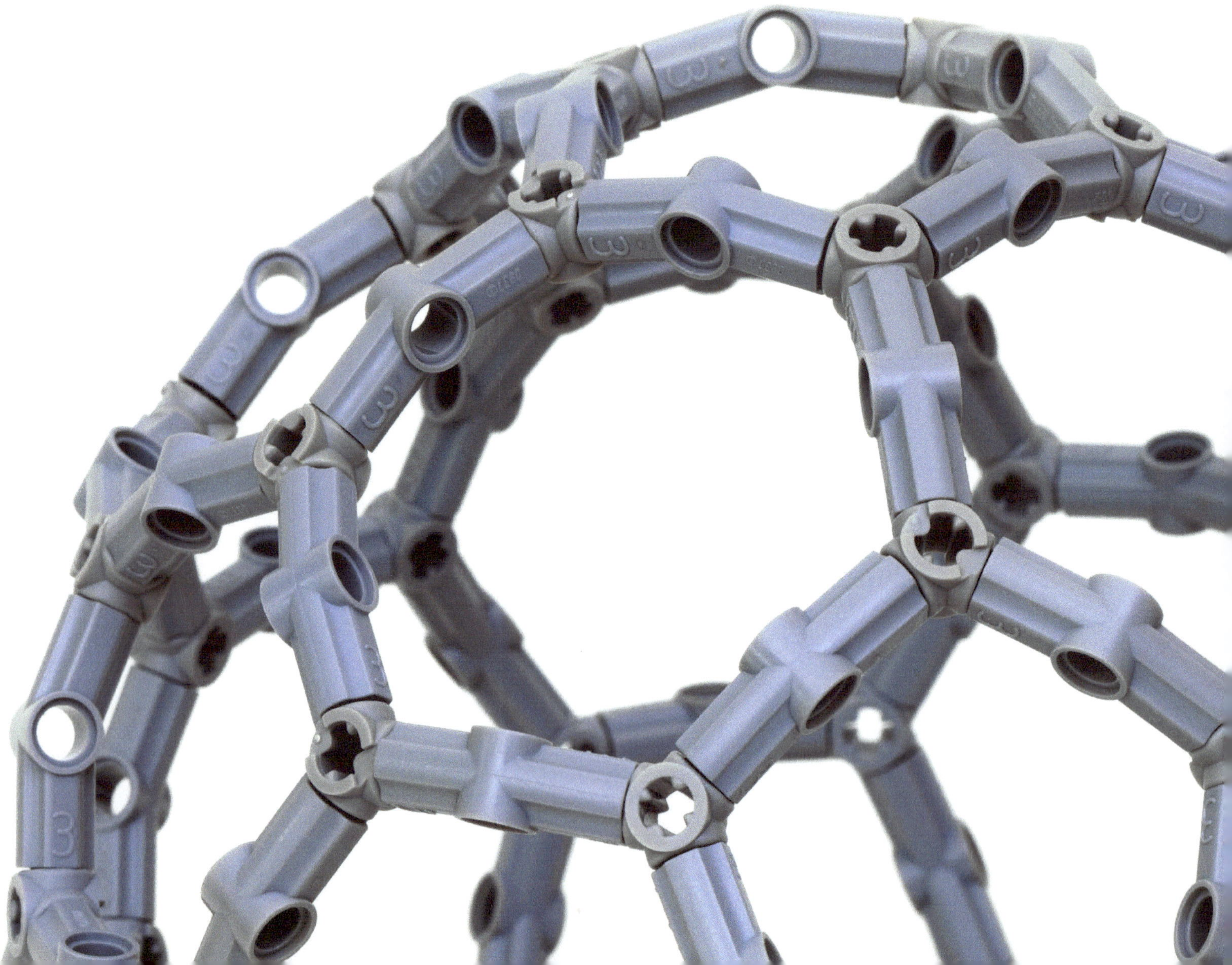

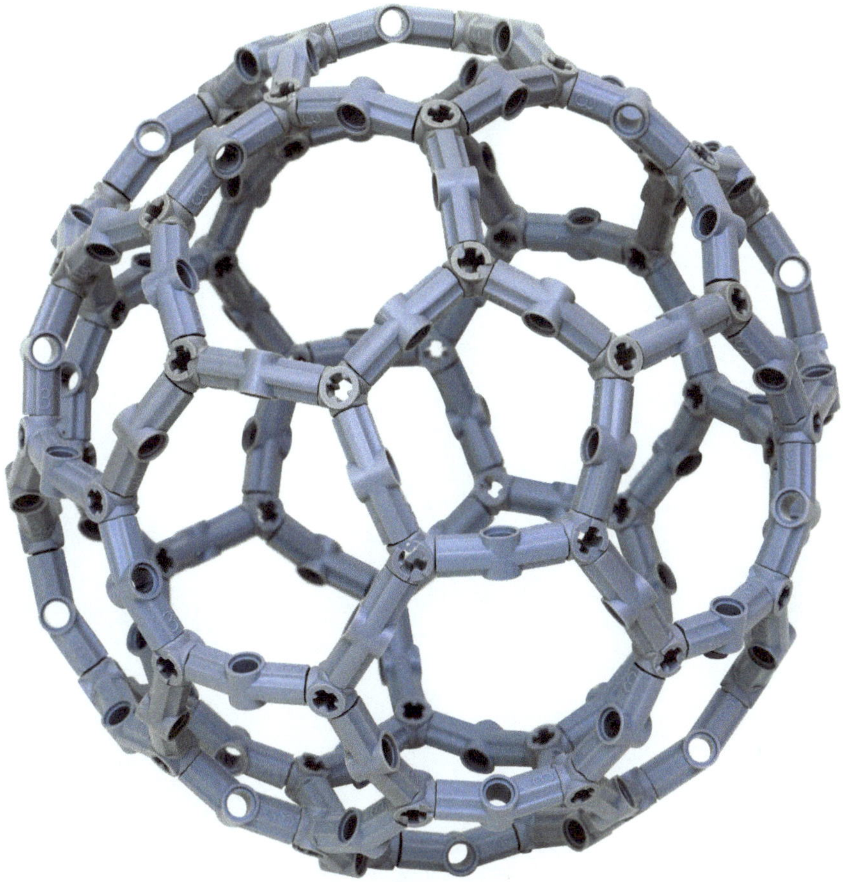

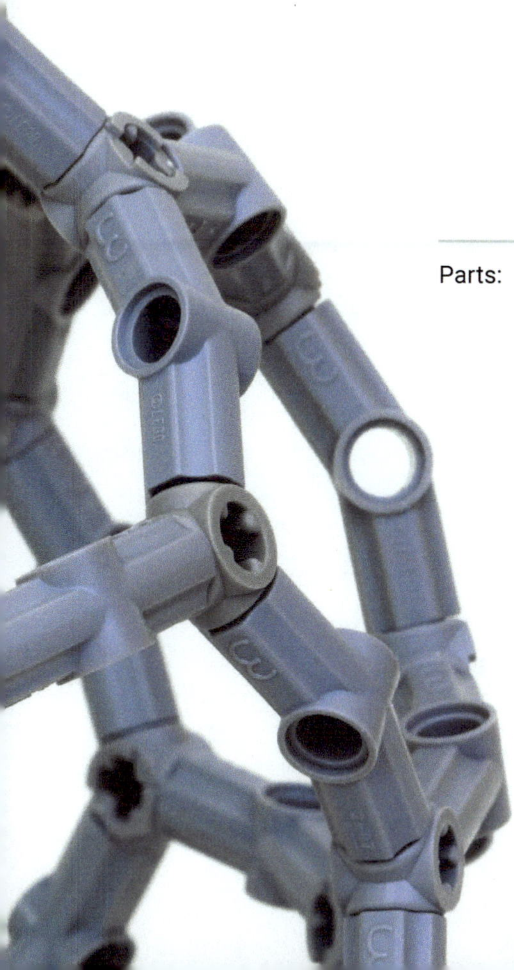

Parts:

3

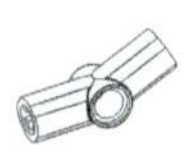

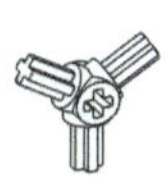

x 90 x 60

EXTRAPOLATIONS

DESIGNS AND VARIATIONS

This section shows other ways of construction using different part types and variations of design. Knots and woven structures are closely related. Joining the open ends of a braid can form knots. Most knots shown here have no open ends and are known as mathematician's knots.

CHAPTER 6

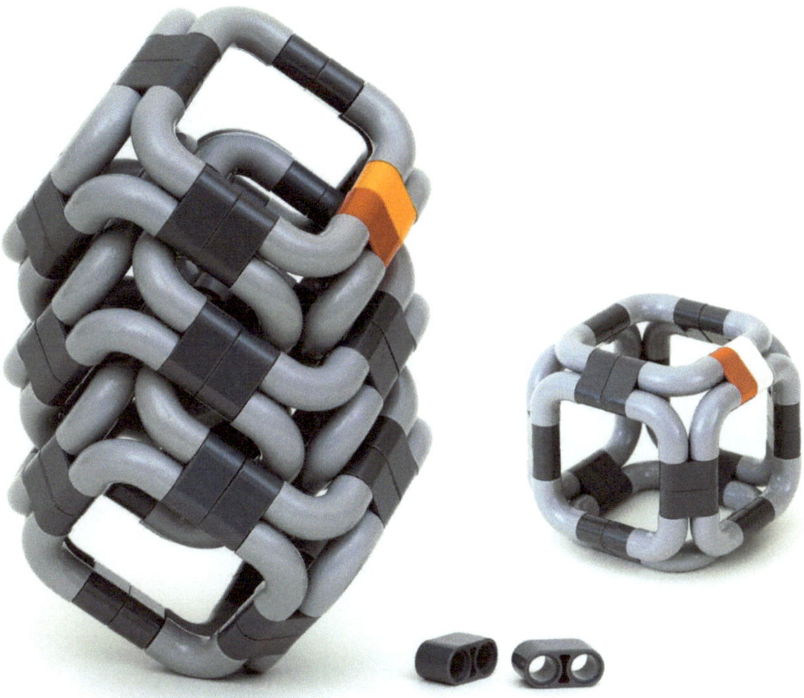

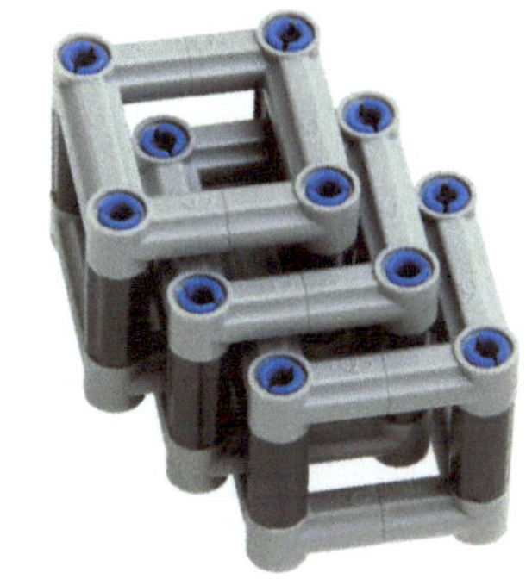

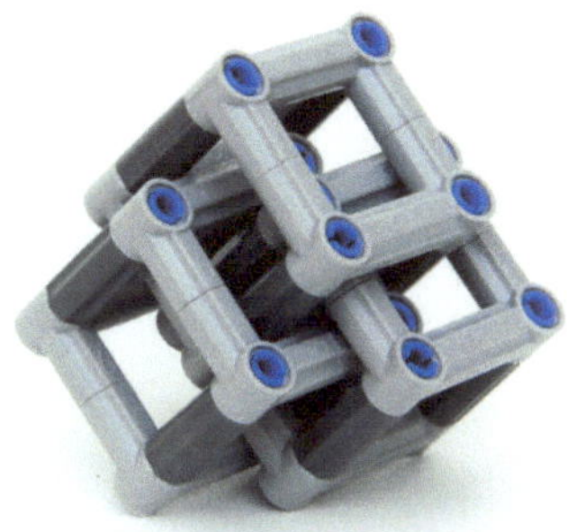

3 nested cubes

3 interlocking rectangles in a Borromean link configuration.

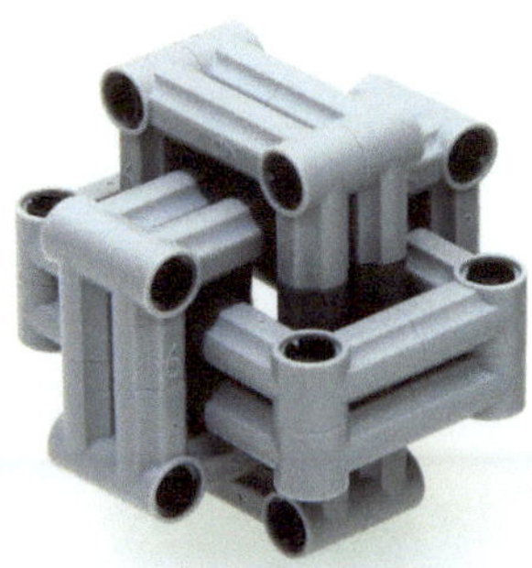

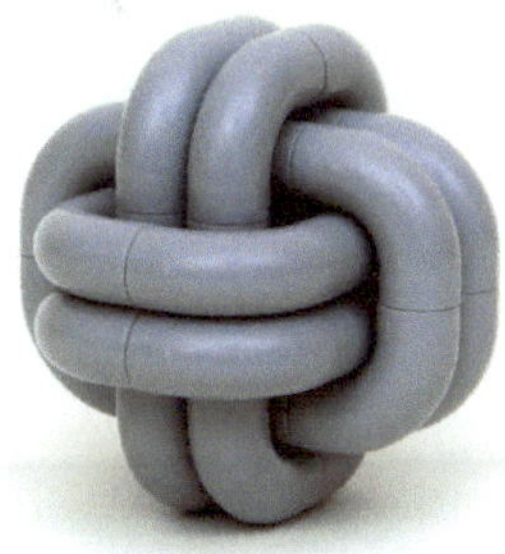

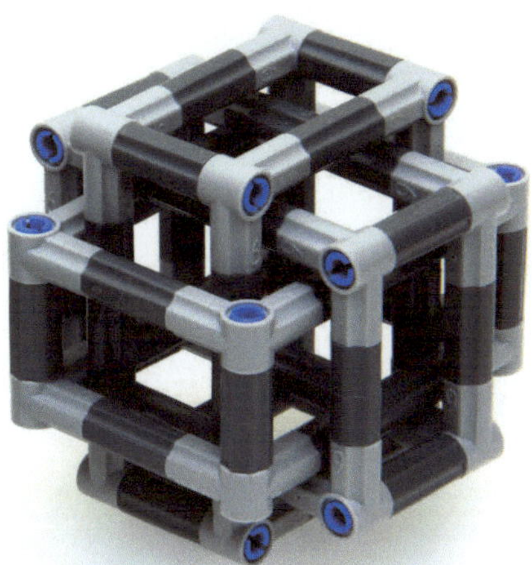

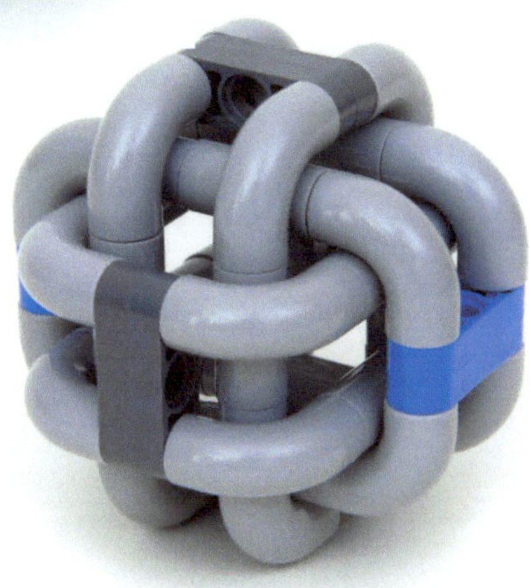

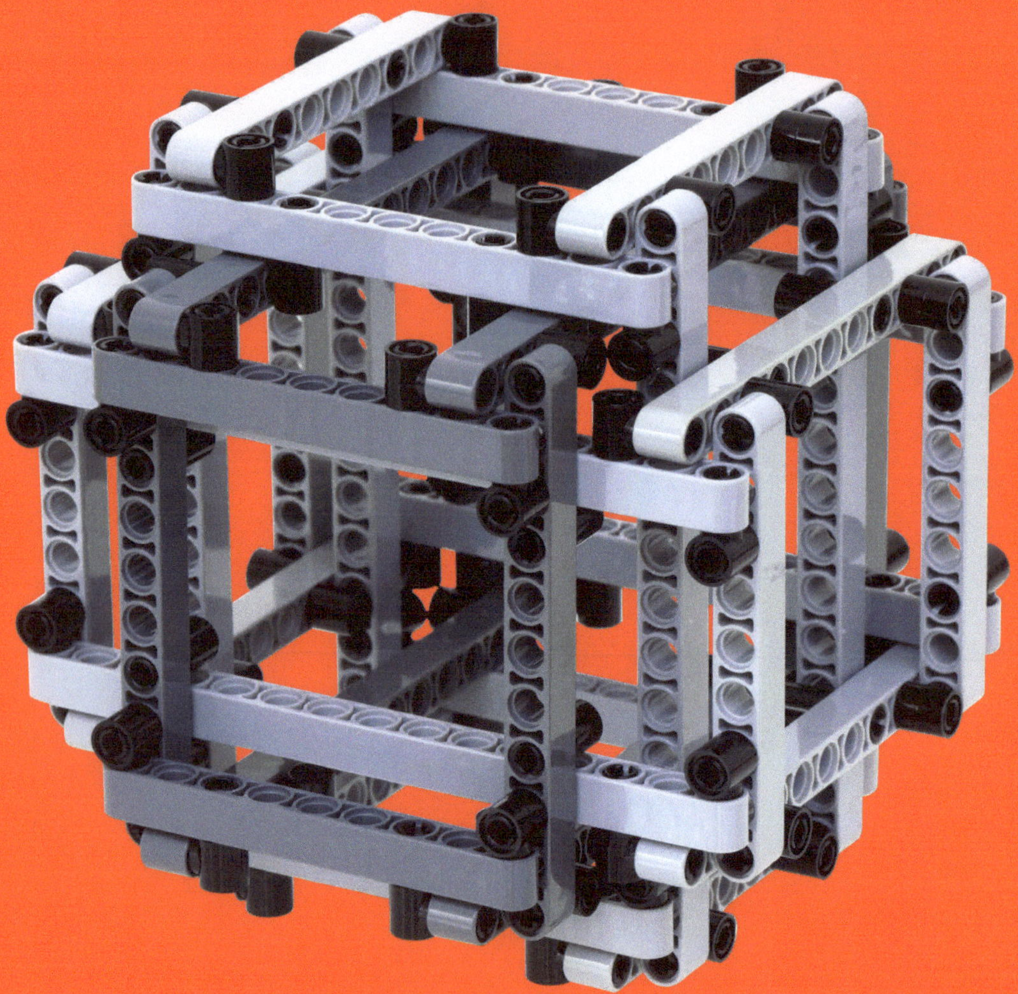

A Cube formed by 3 rectangular prisms in a Borromean link configuration.

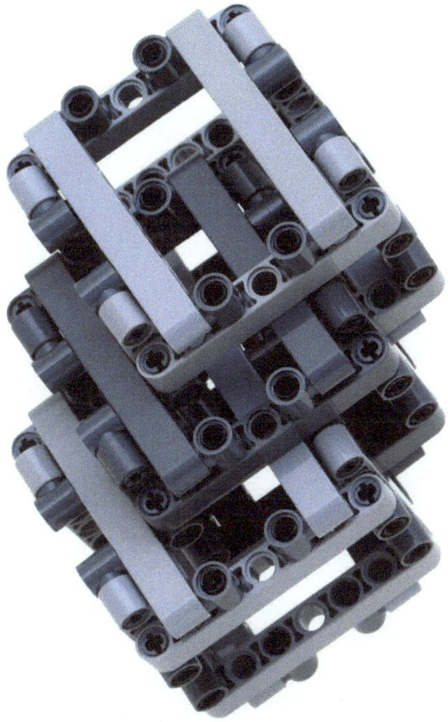

3 nested cubes

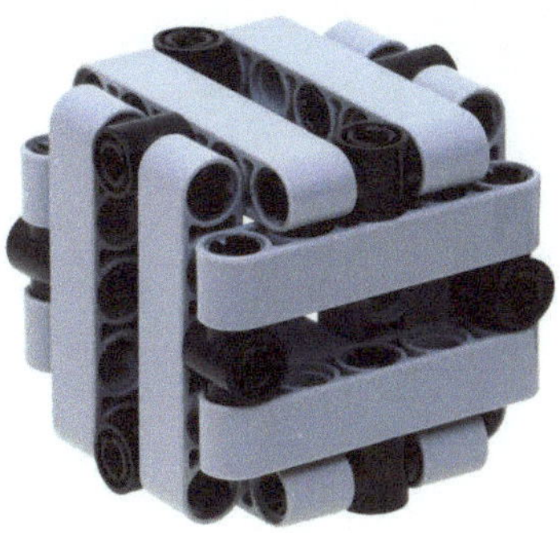

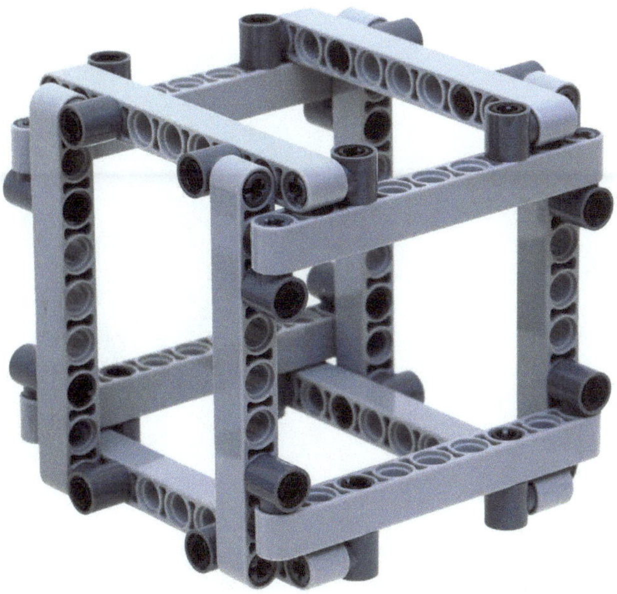

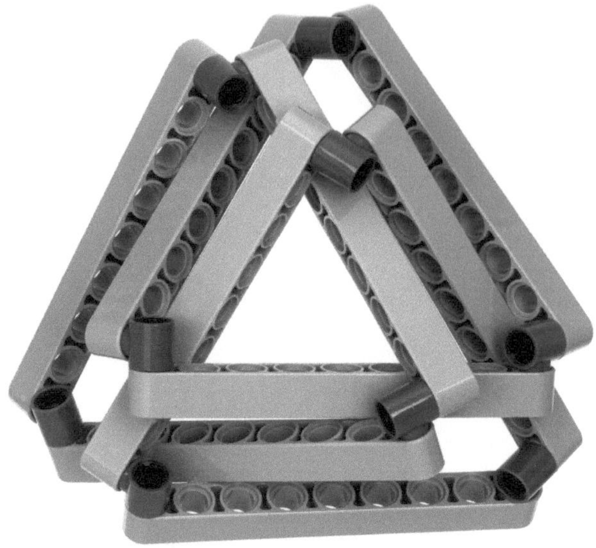

3 Triangles in a Borromean weave configuration.

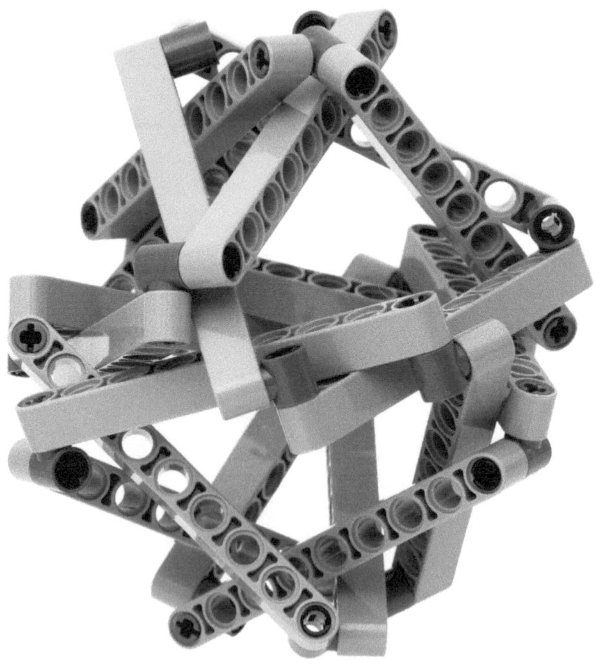

2 sets of the above assembly intersecting to form 6 triangles.

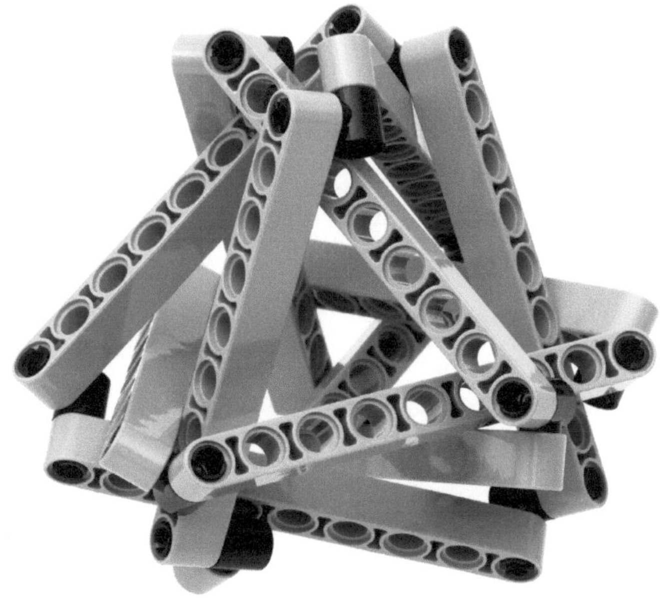

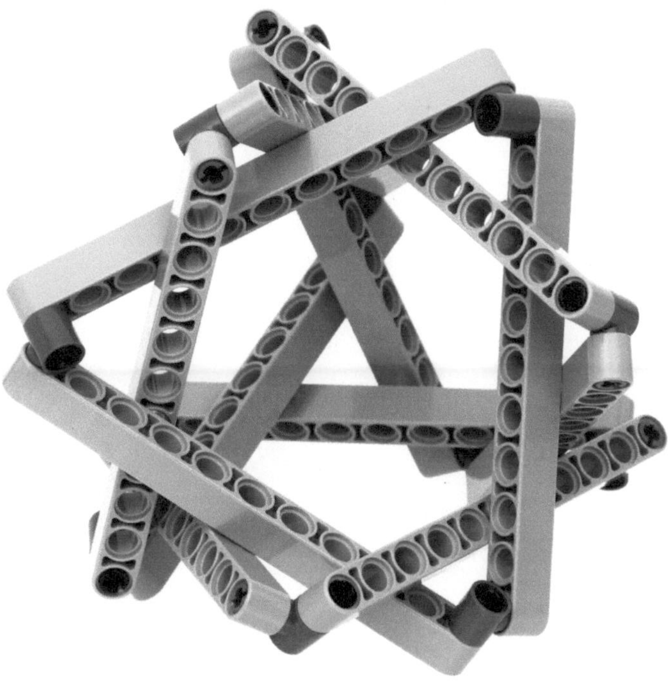

4 intersecting triangles - 2 versions.

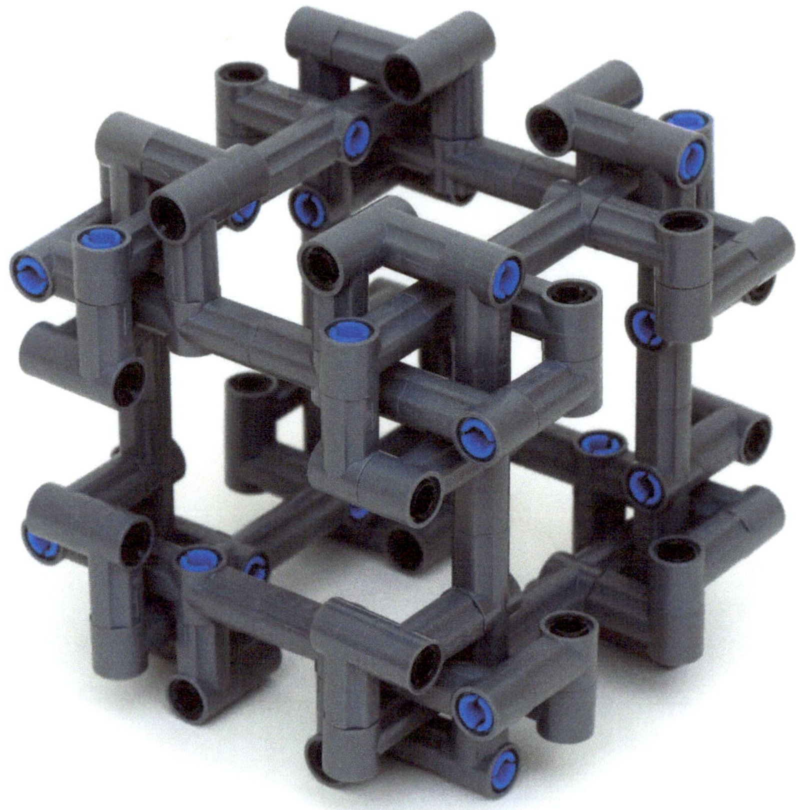

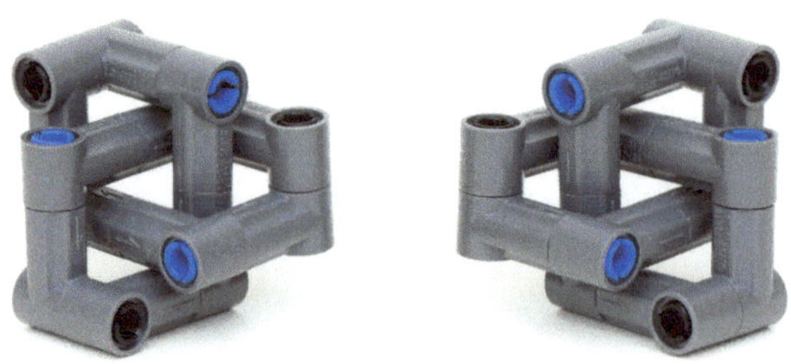

A pair of left-hand and right-hand trefoil knots used as cube corners of this cube.

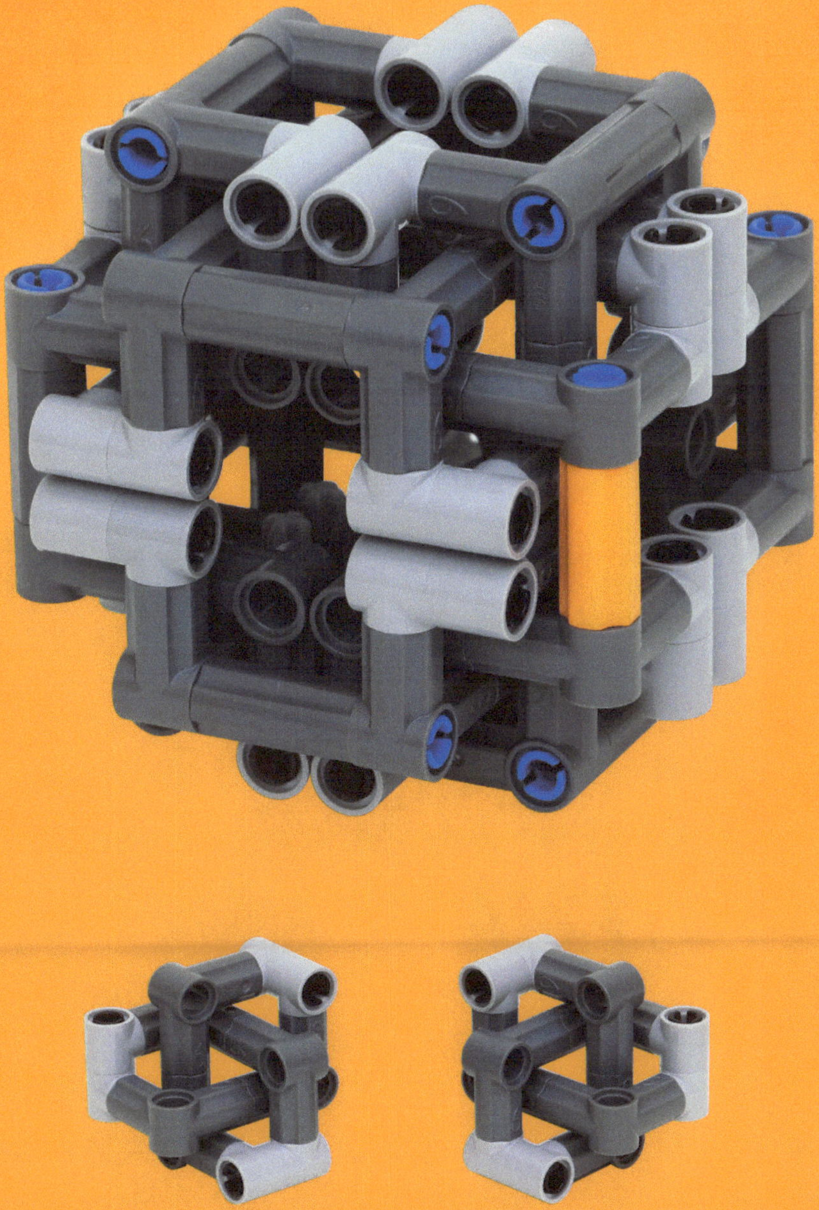

Another variation of trefoil cube corners.

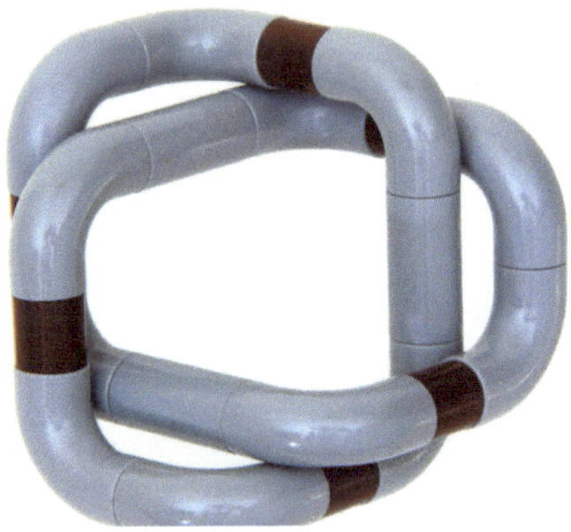

A Trefoil knot is an overhand knot joined at the ends.
It is the simplest Gordian knot.

These cubes' corners are formed by trefoil knots.

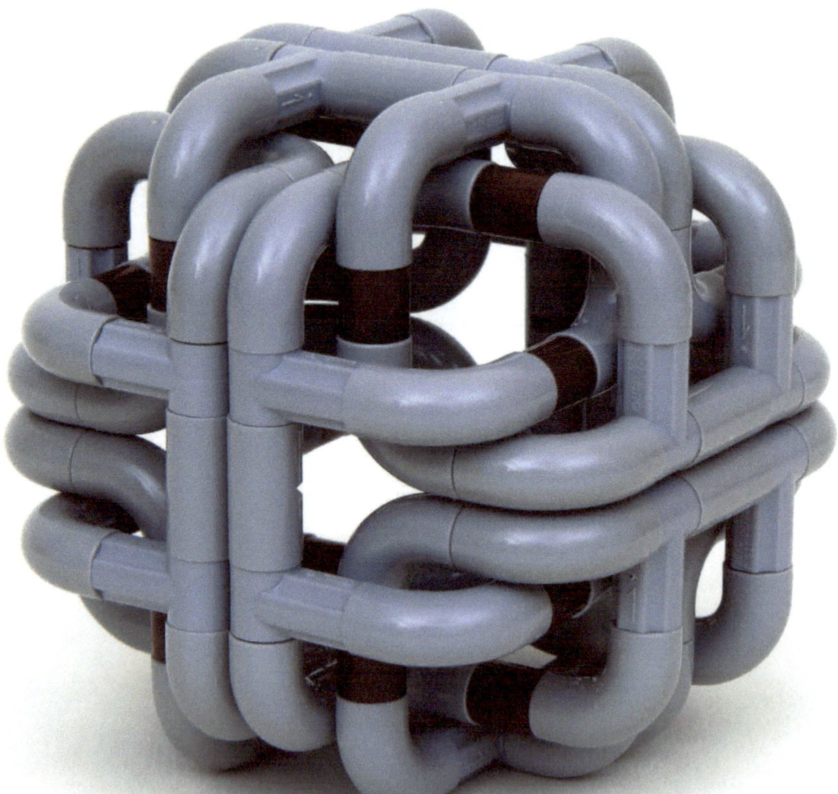

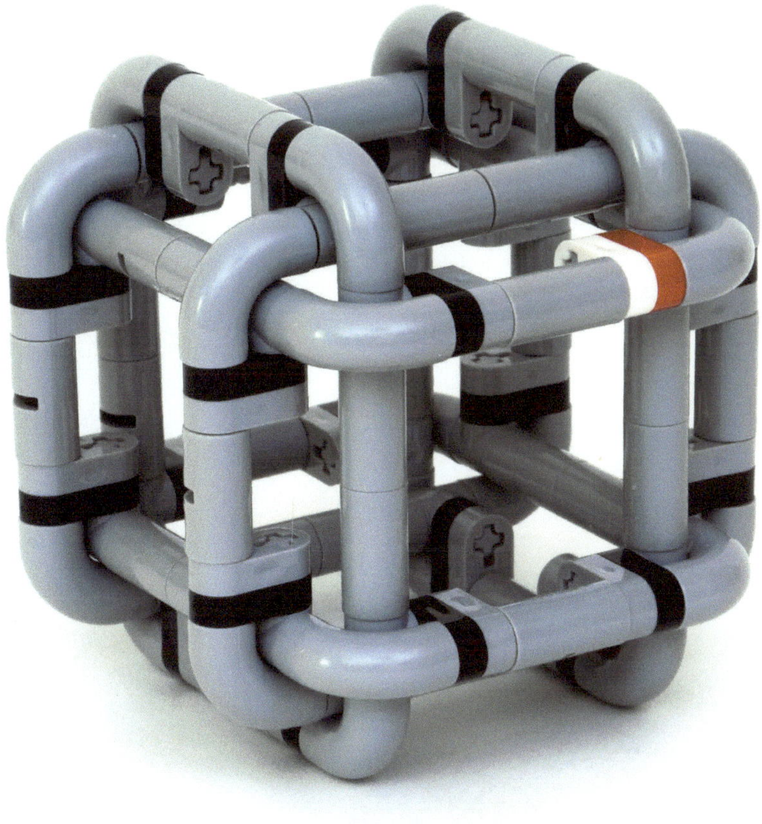

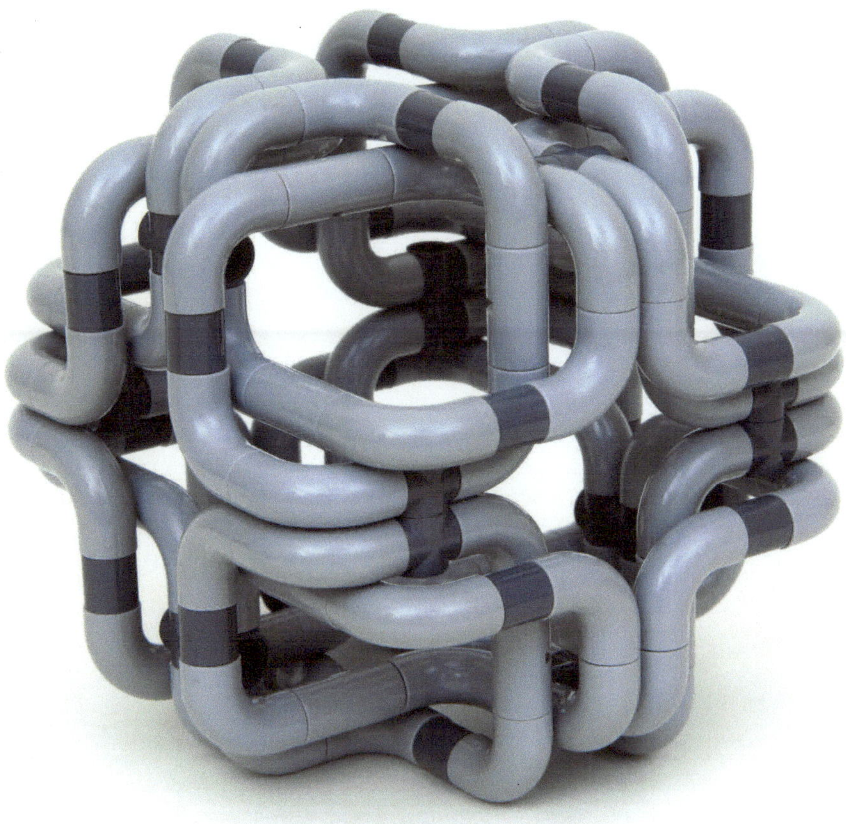

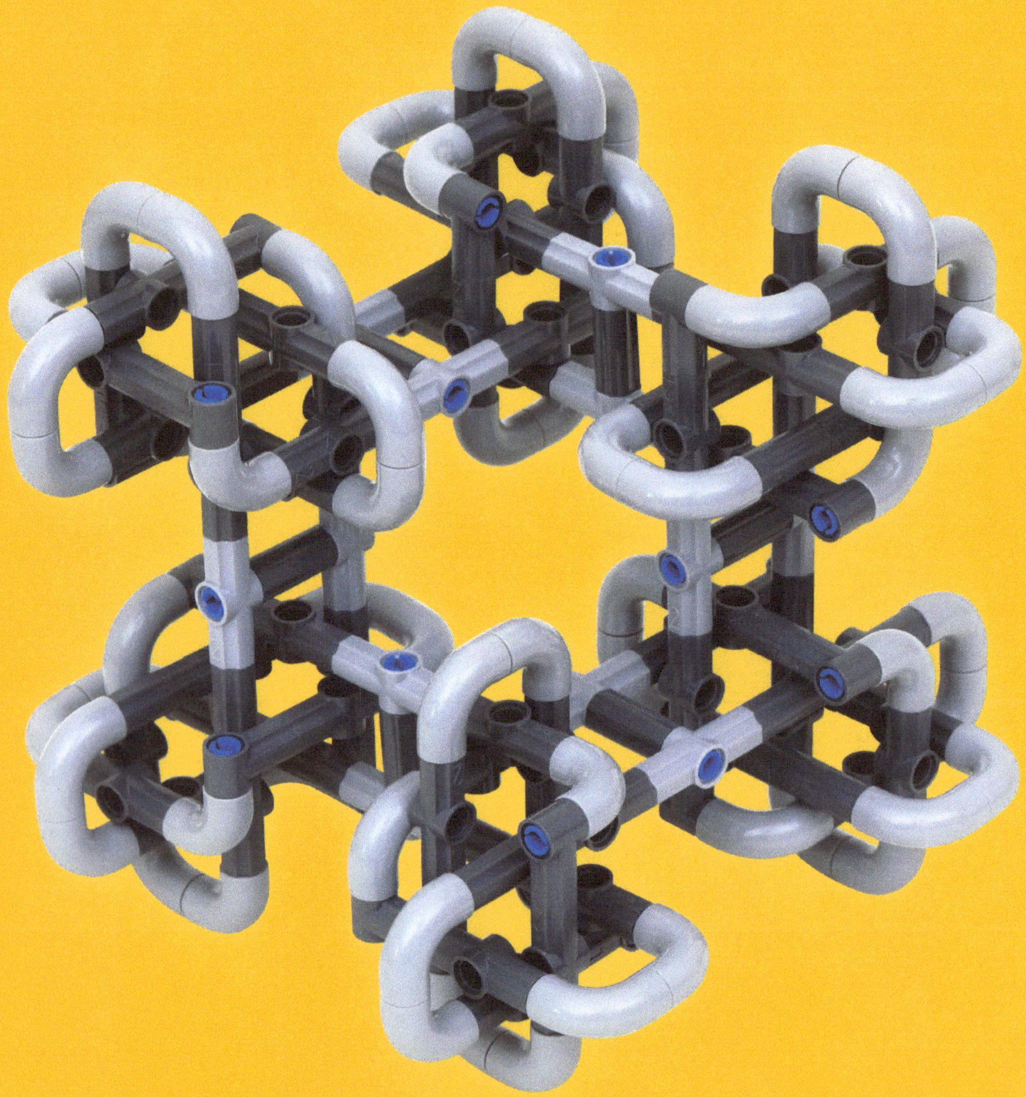

6 connected endless knots.

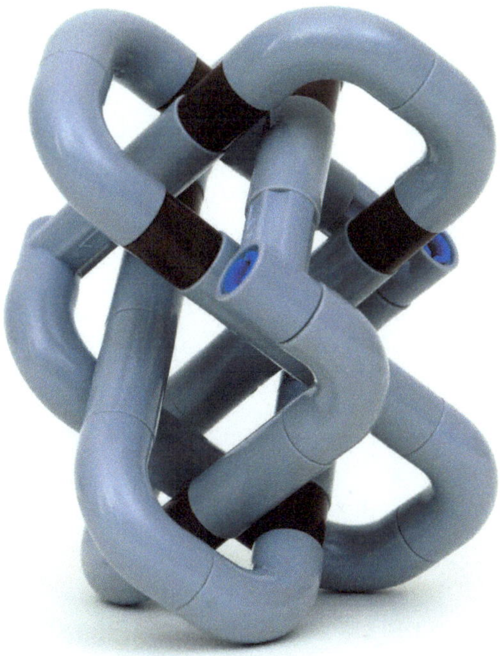

This is similar to the endless knot and the Chinese button knot.

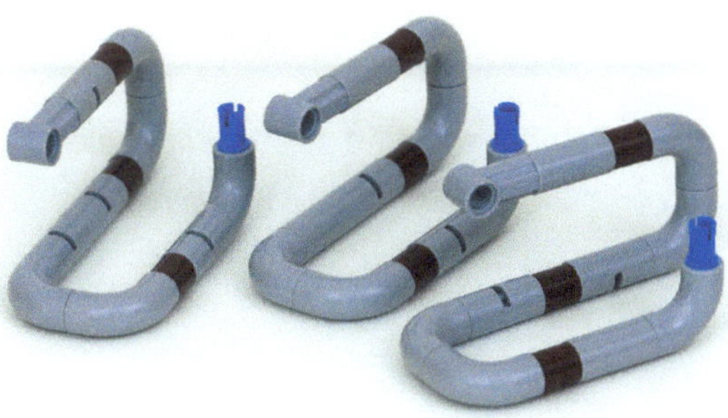

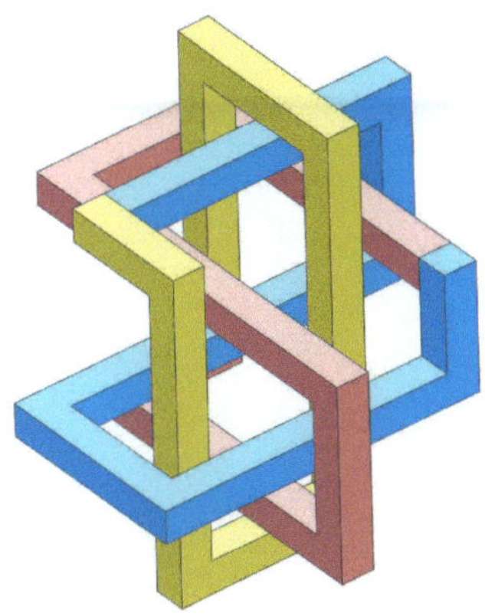

It is made of 3 identical pieces and assembled orthogonally to each other.

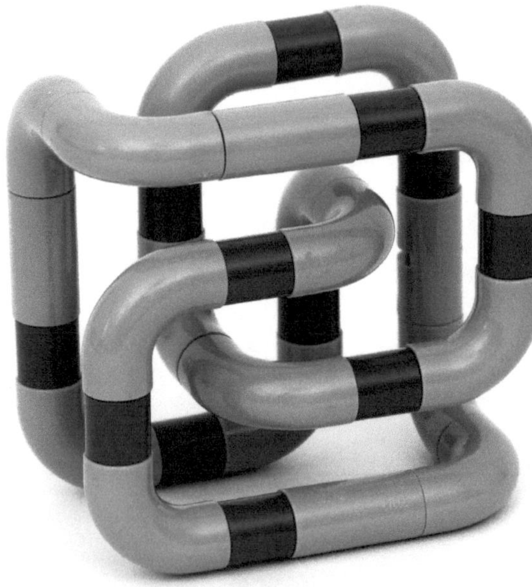

A trefoil knot variation.

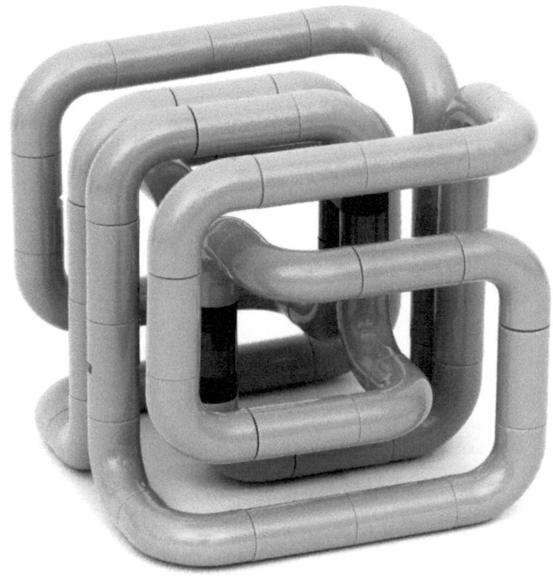

A left-hand and right-hand pair of trefoil knots forming a cube structure.

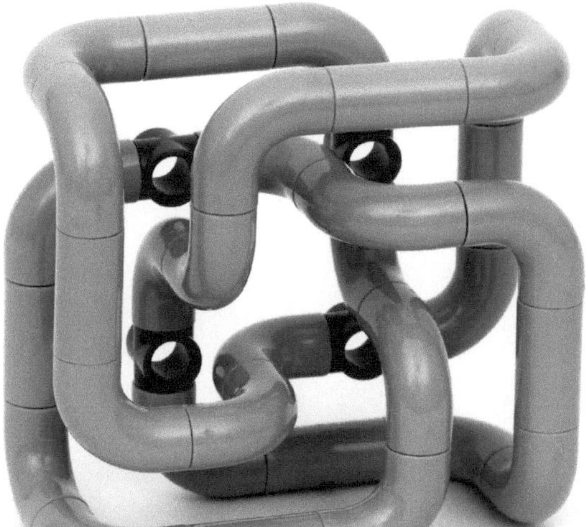

A Quatrefoil knot.

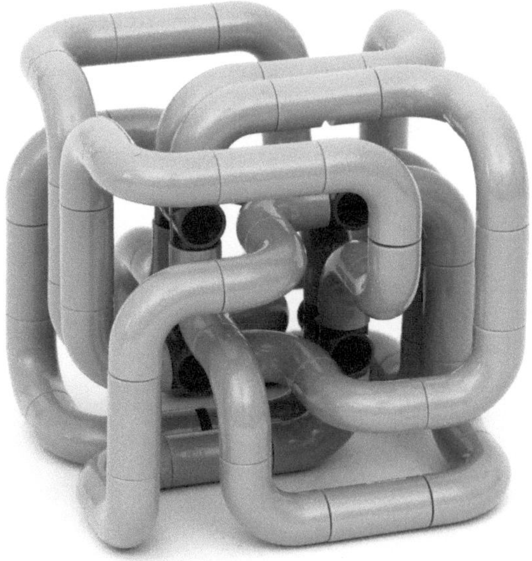

Left-hand and right-hand pair forming a cube.

A trefoil knot using 90 degree elbows. This knot is visually similar to the triquetra which is also known as a trinity knot.

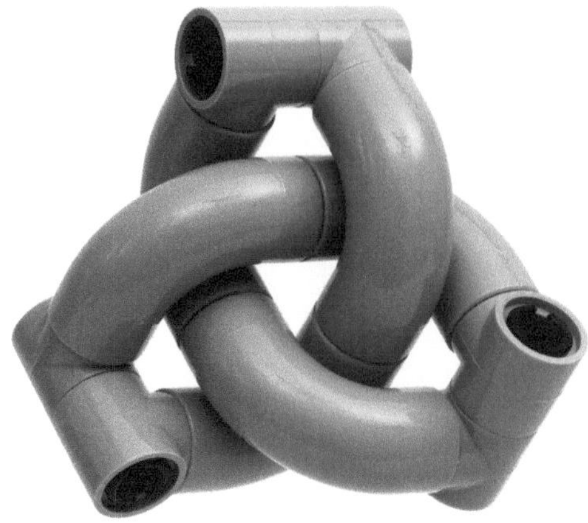

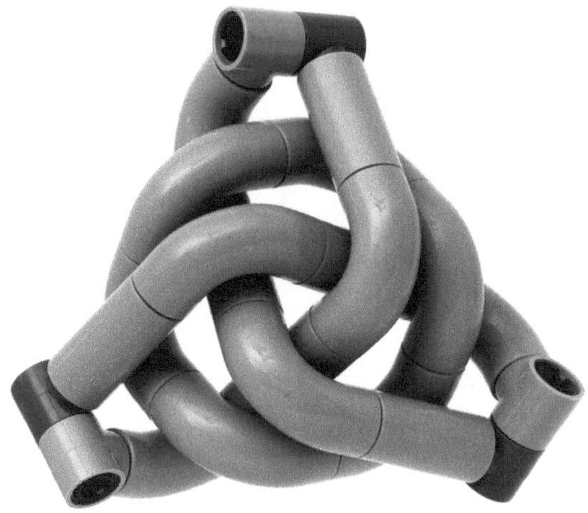

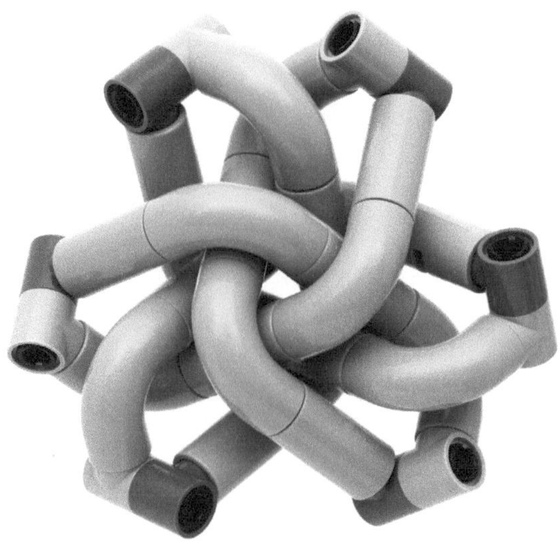

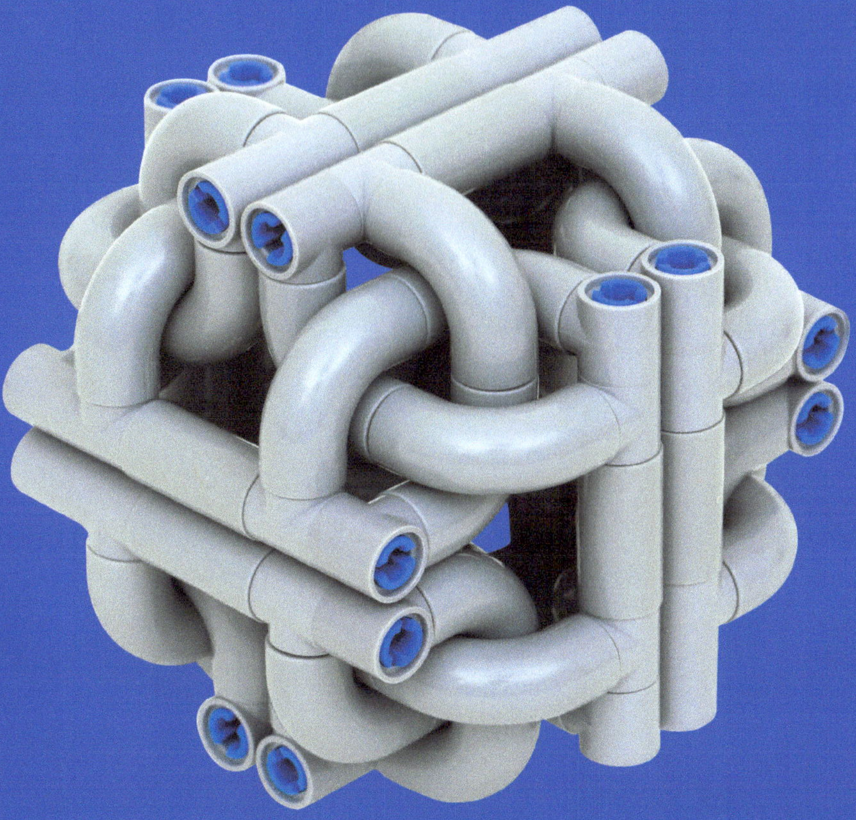

A cube based on this knot.

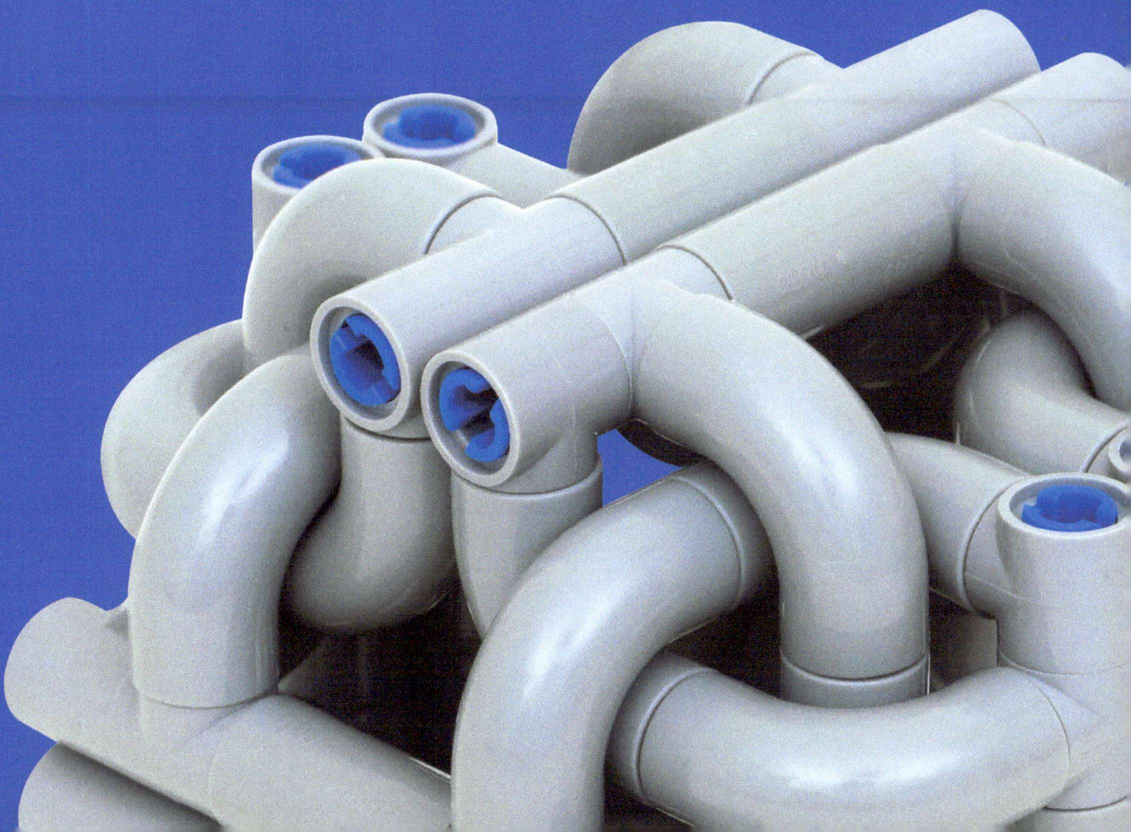

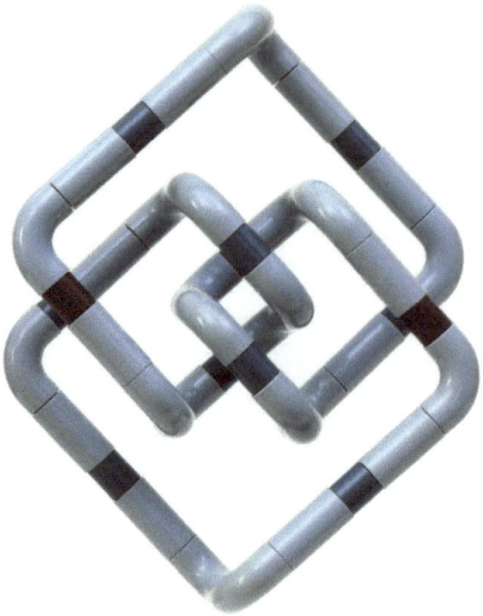

Listing's Knot also known as Figure Eight Knot. A symmetric version.

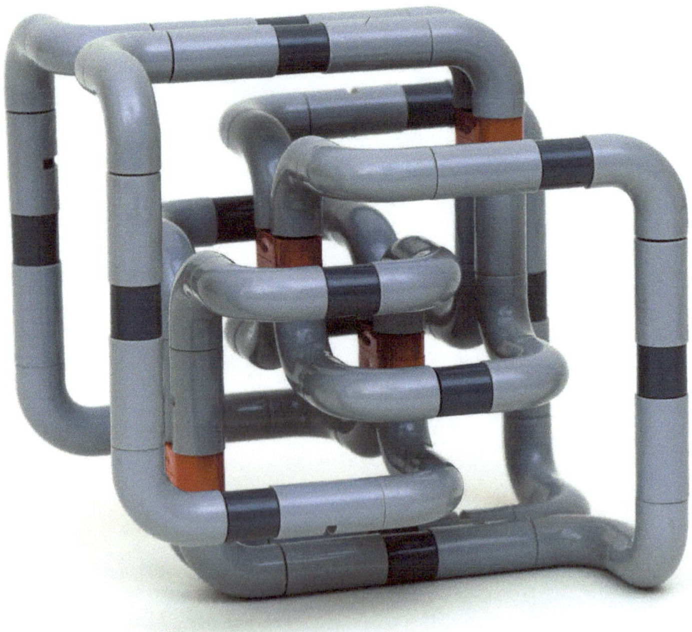

A pair of right-hand and left-hand Listing's knot.

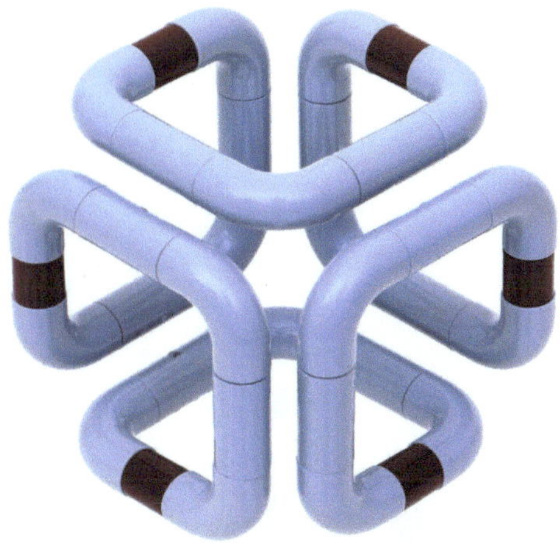

Infinity cube.

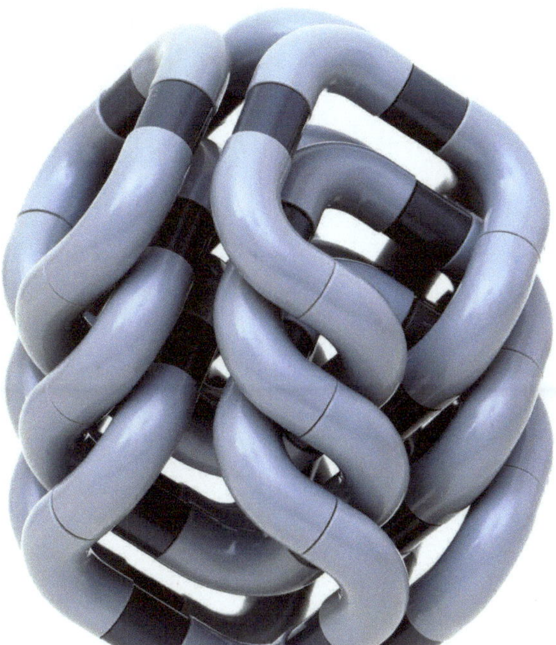

3 nested Infinity Cubes.

WOVEN And Repeating Structures

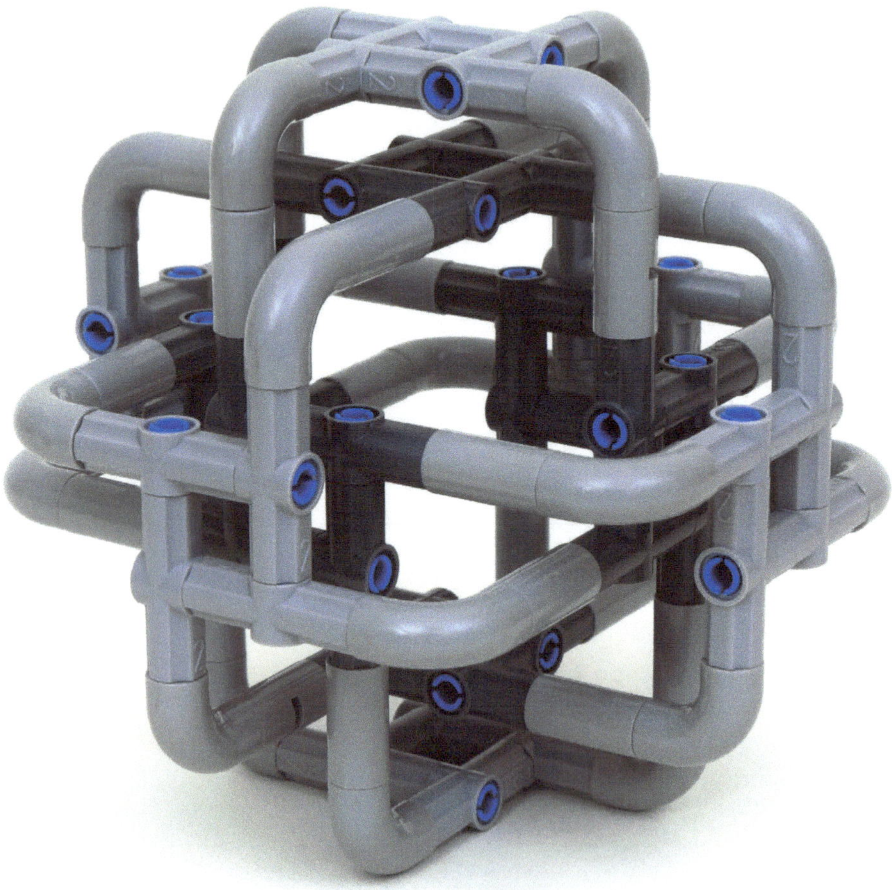

A woven cube.

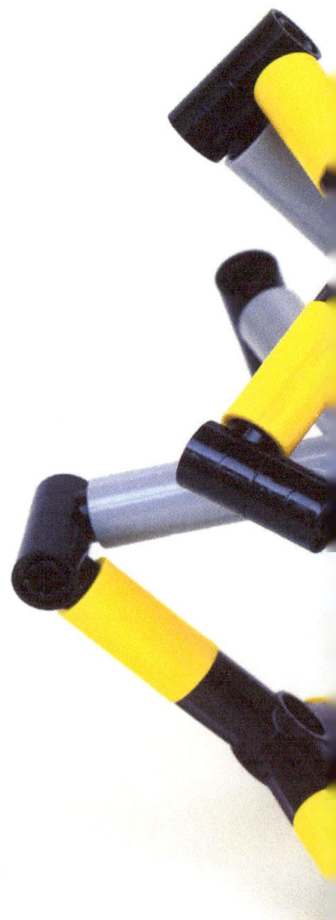

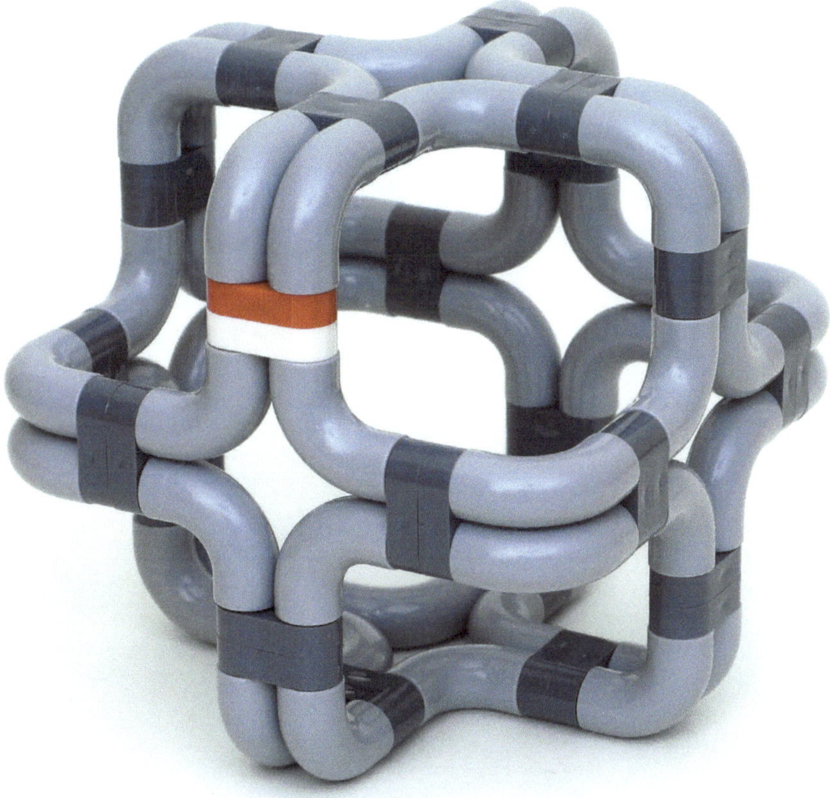

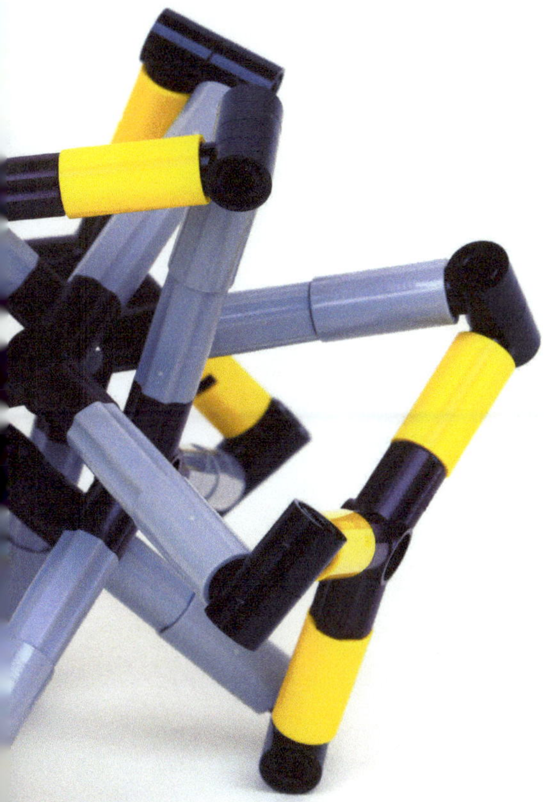

A woven structure based on the tetrahedral configuration.

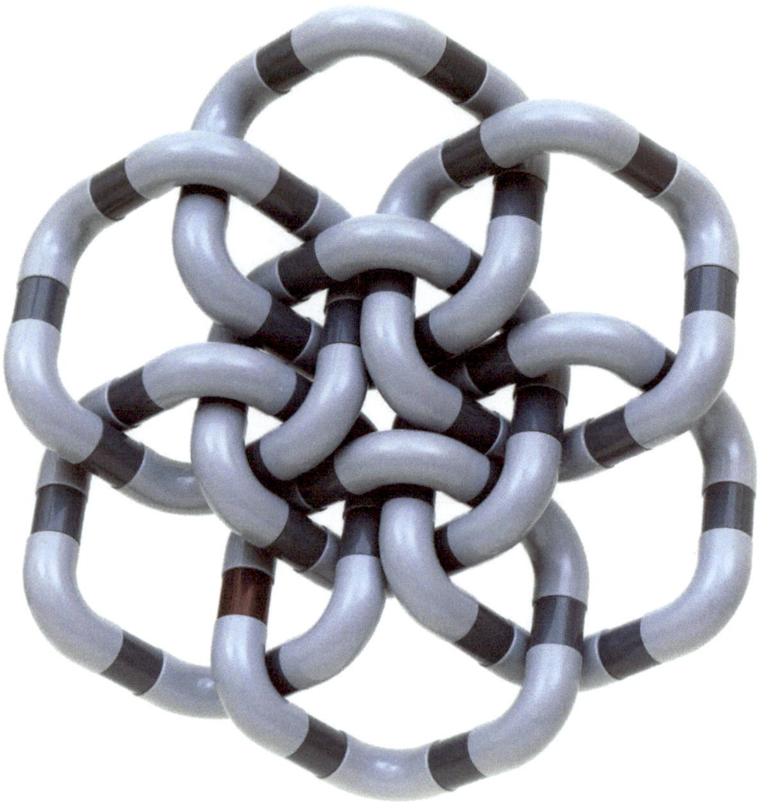

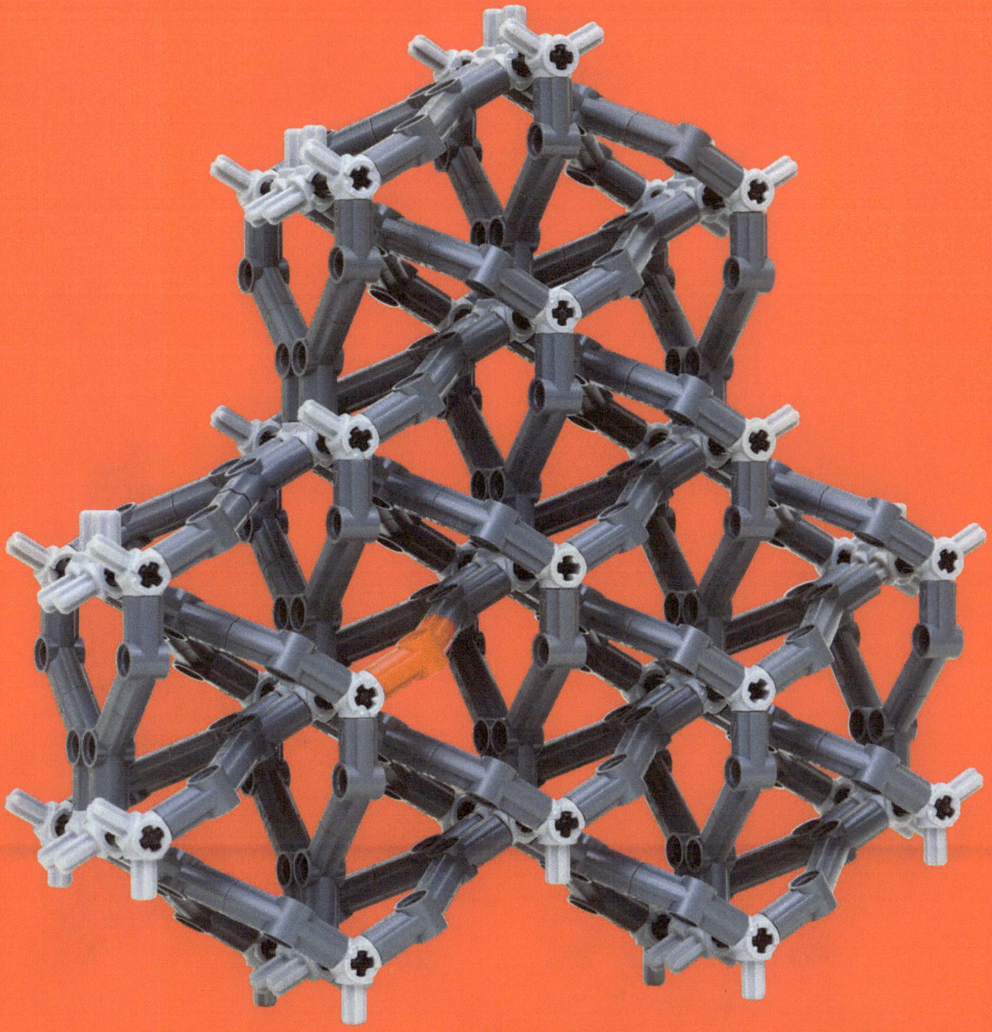

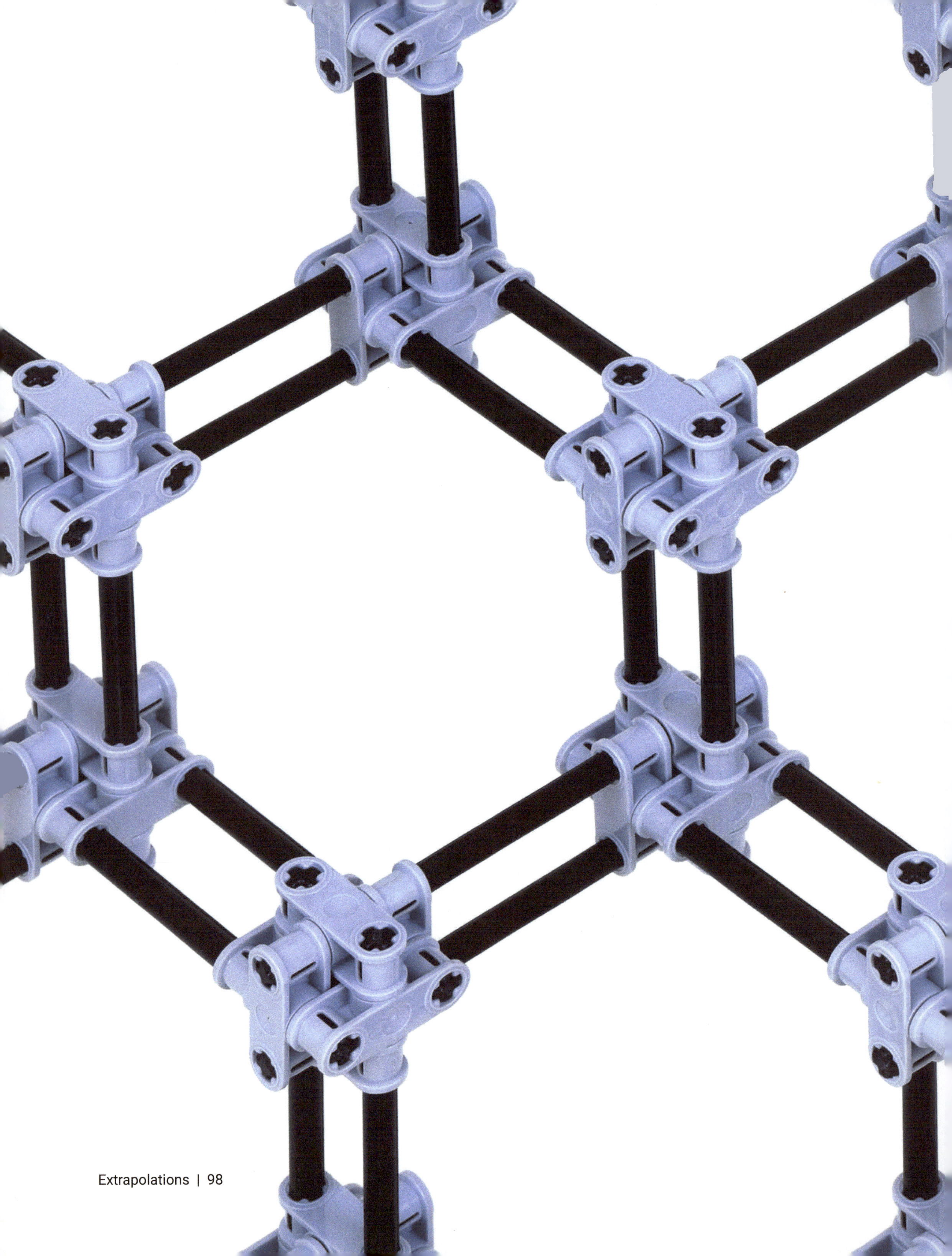

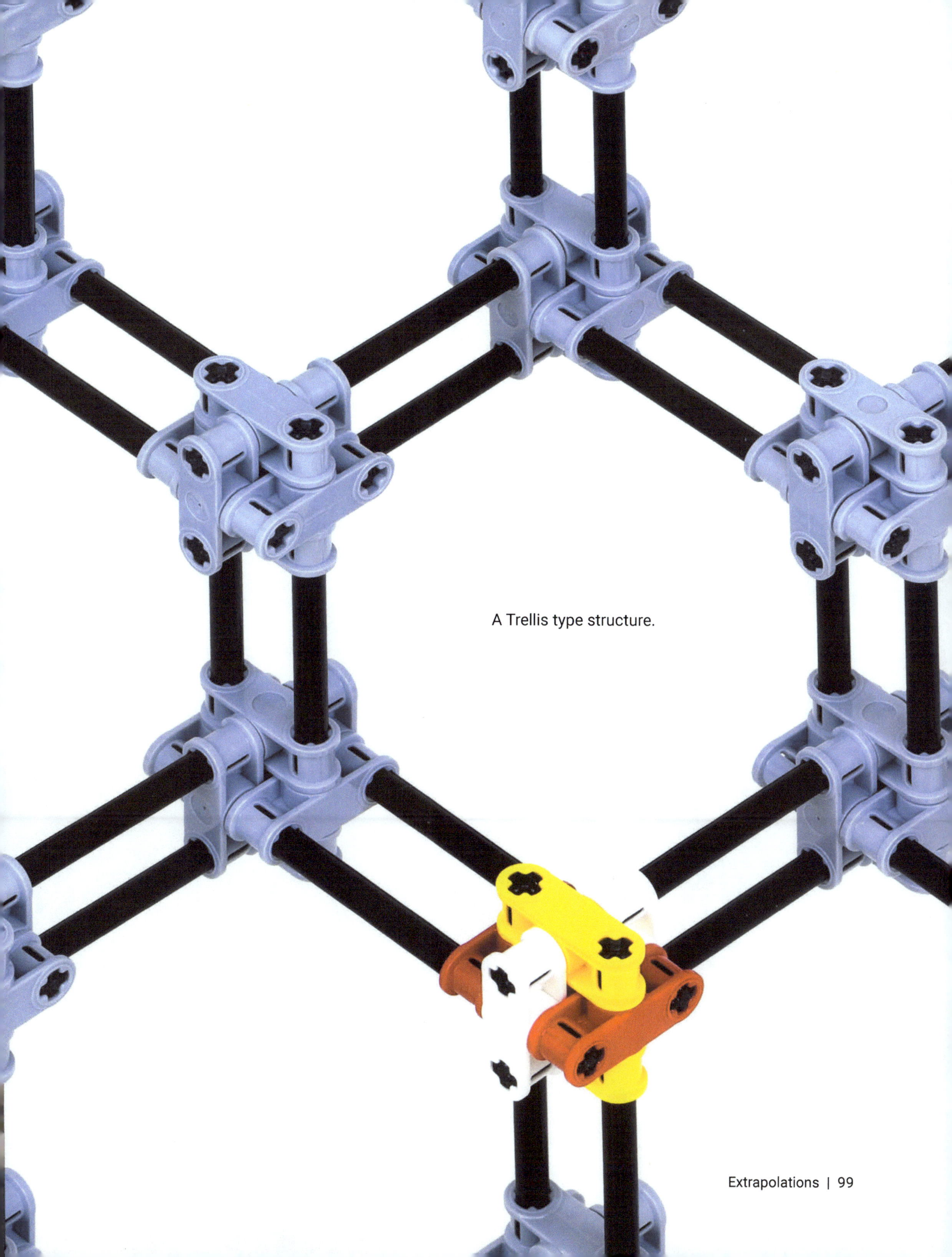

A Trellis type structure.

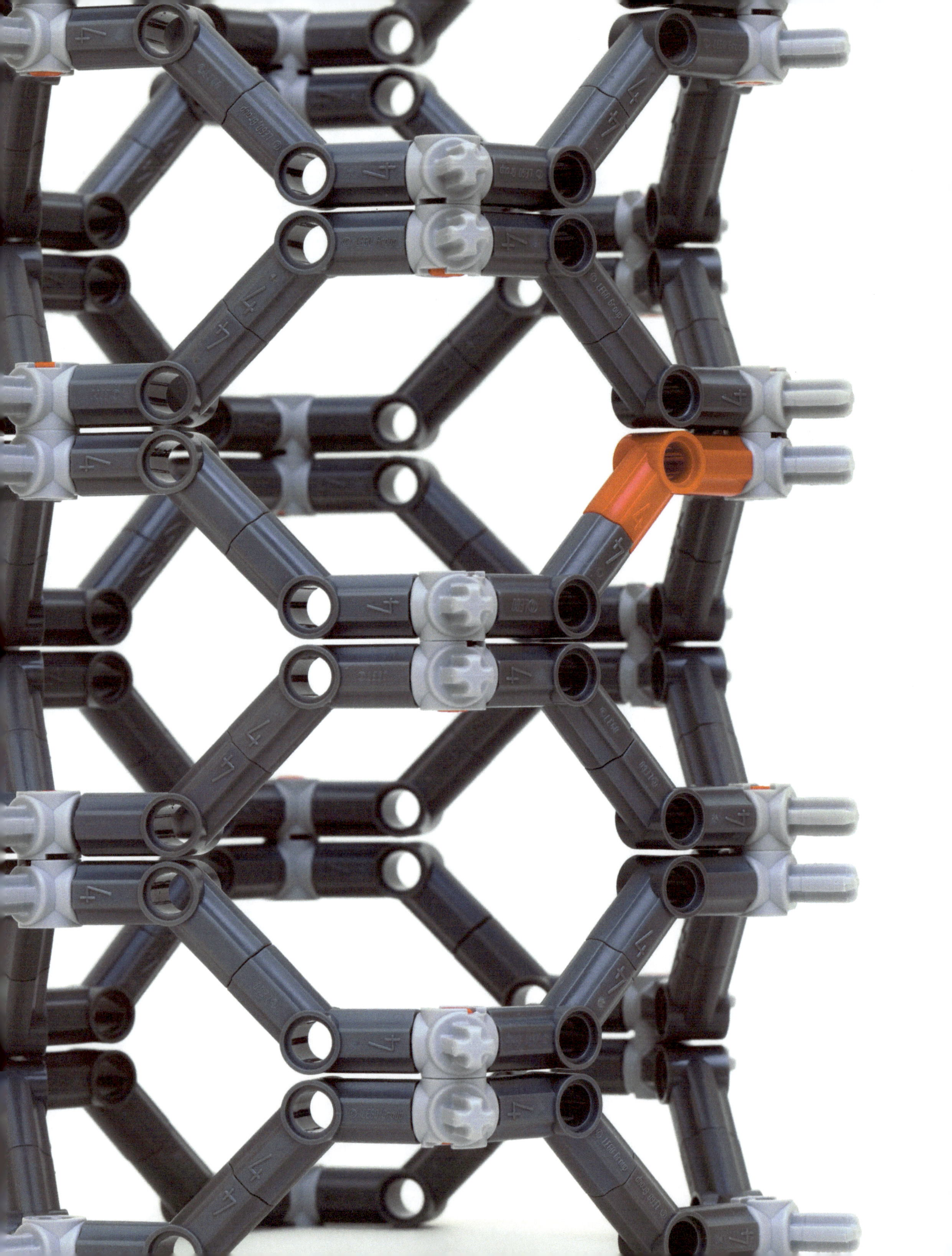

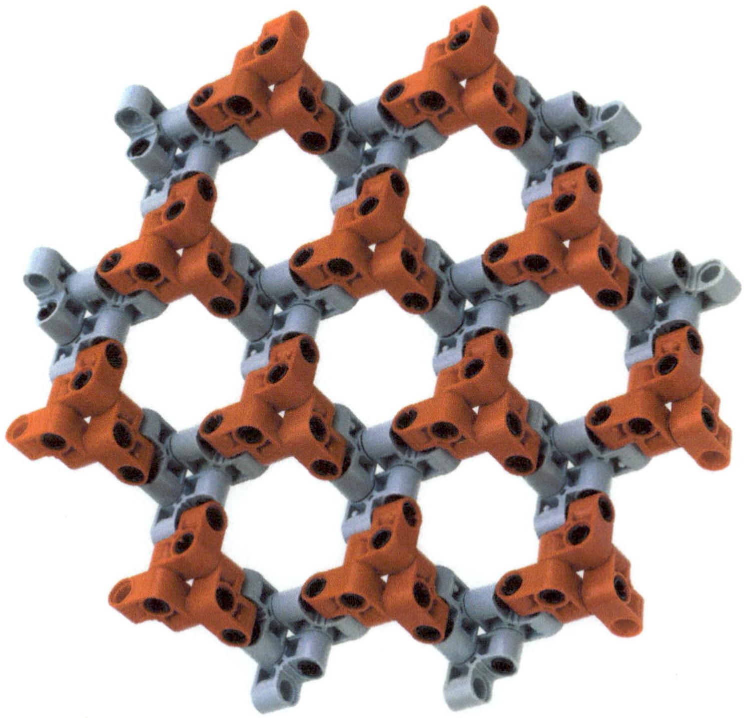

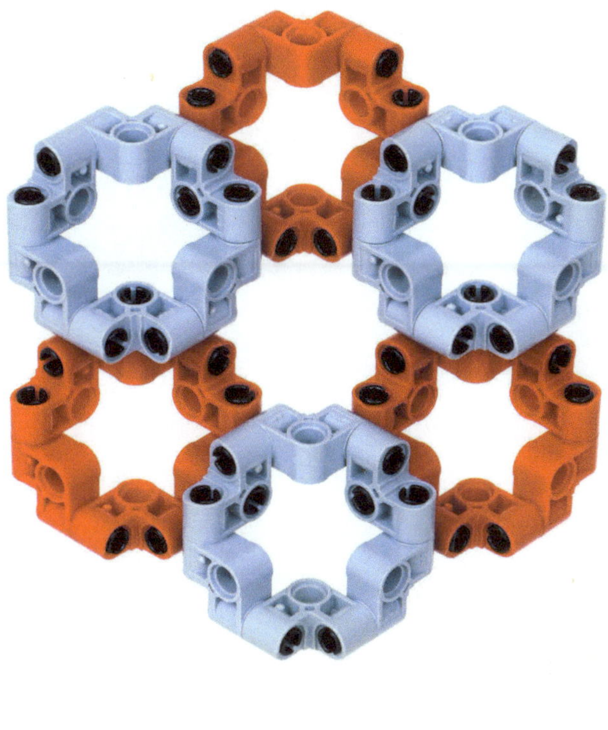

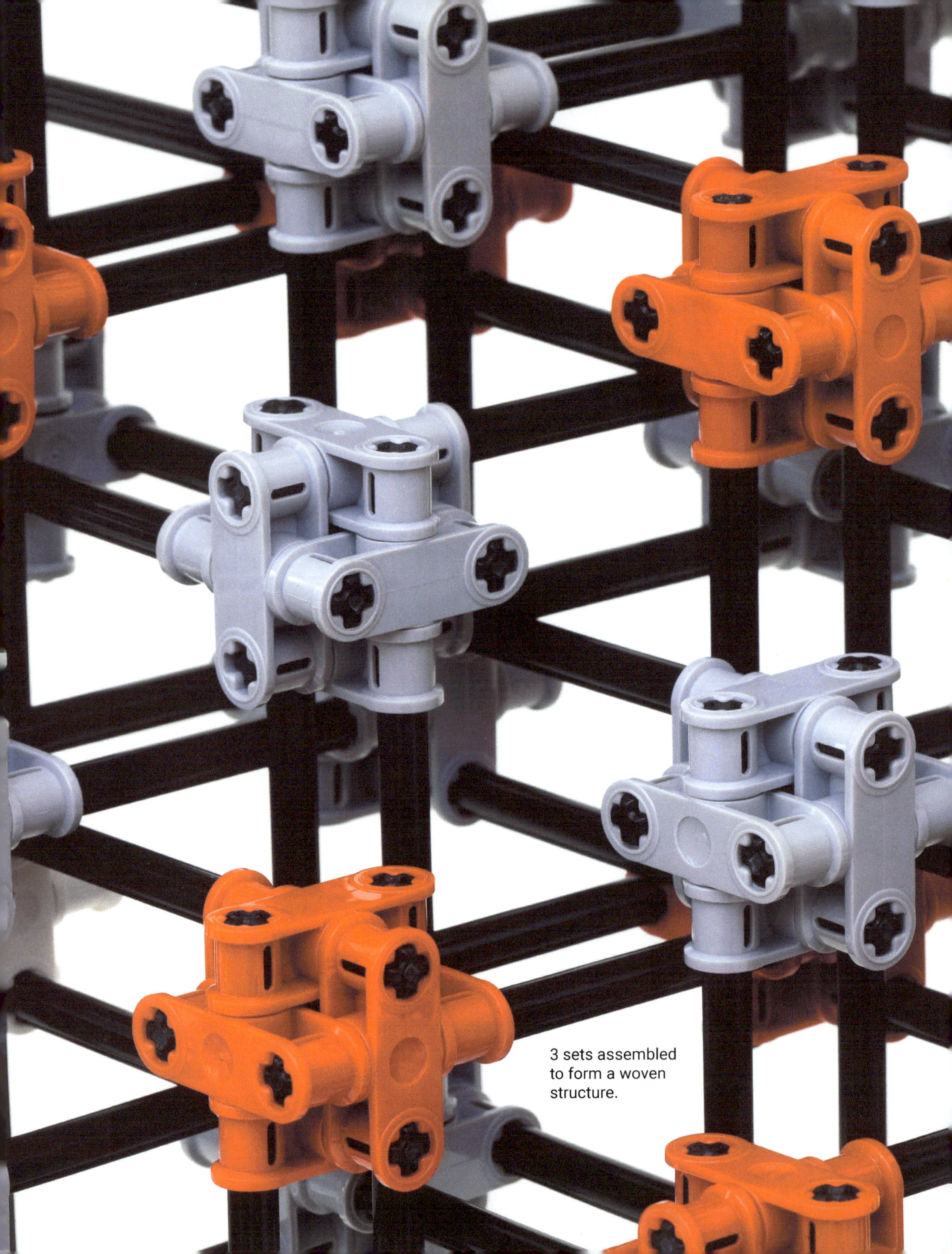

3 sets assembled to form a woven structure.

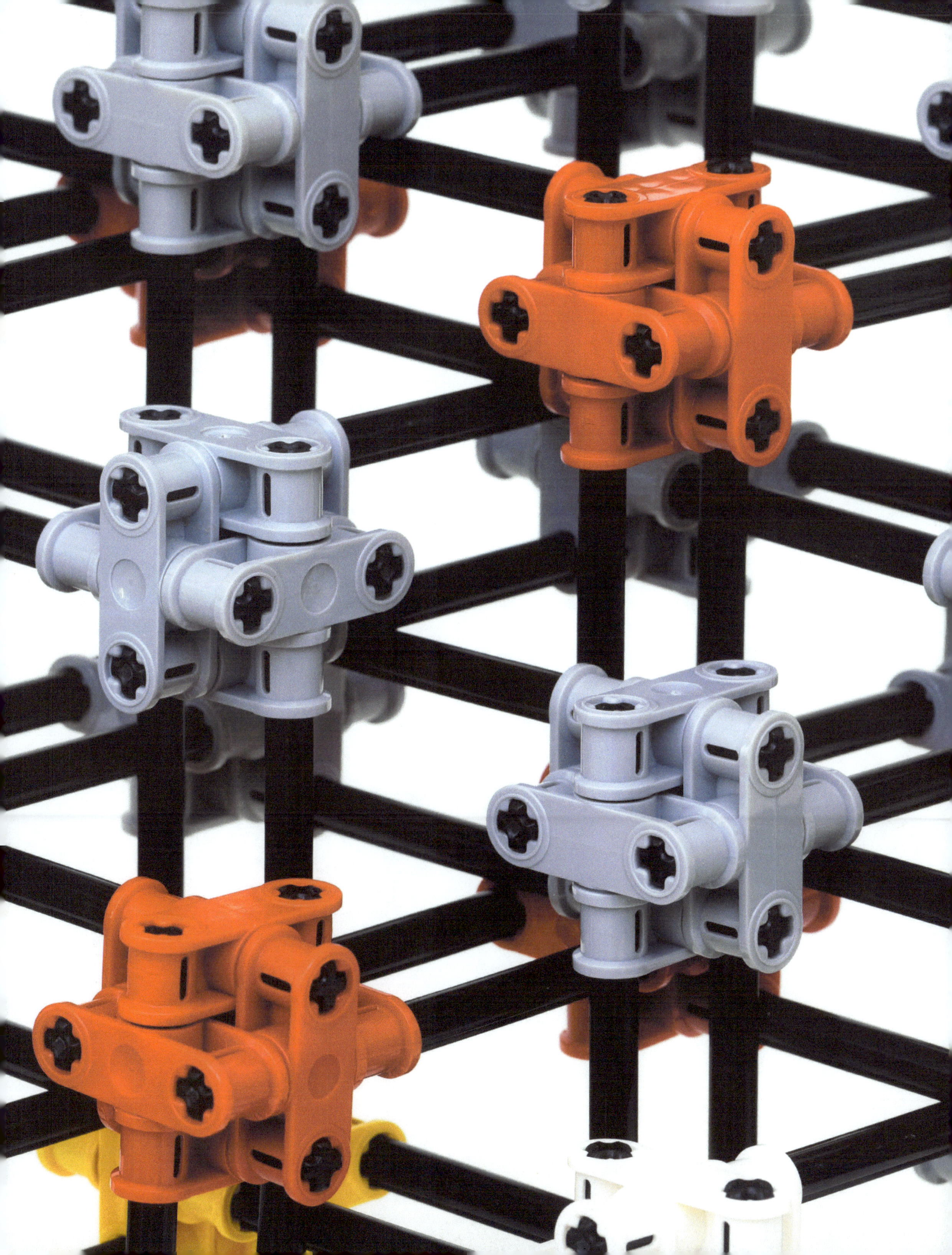

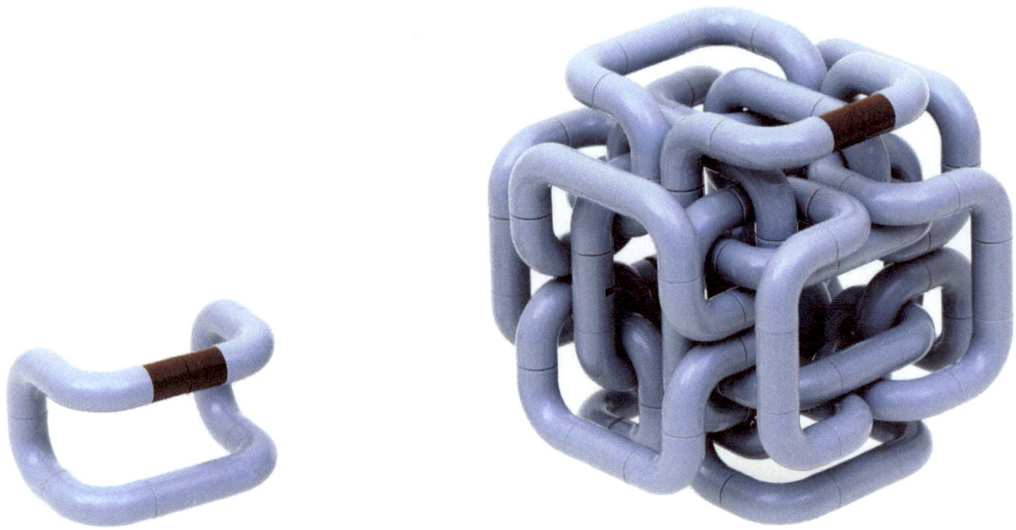

A Cube formed by 12 identical links in a Borromean weave configuration.

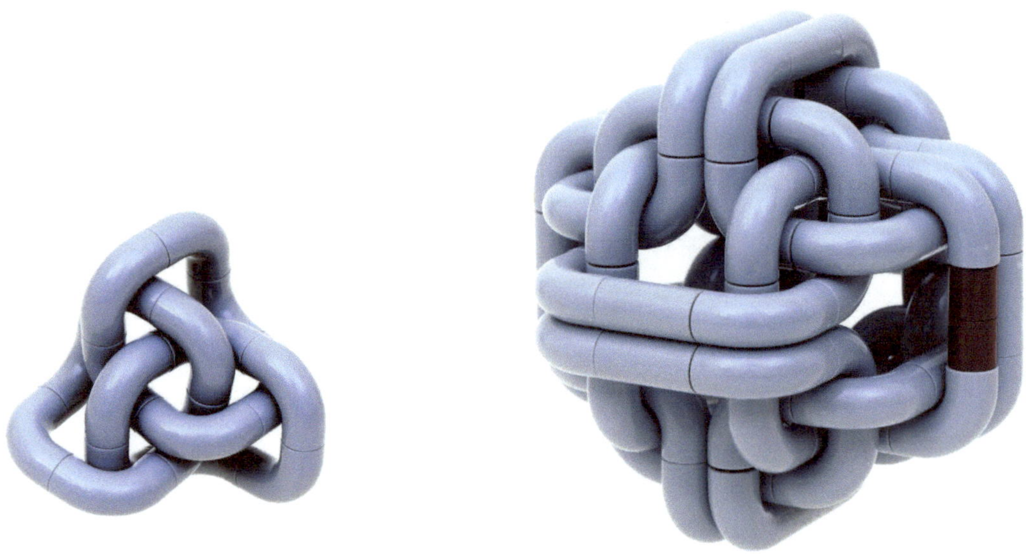

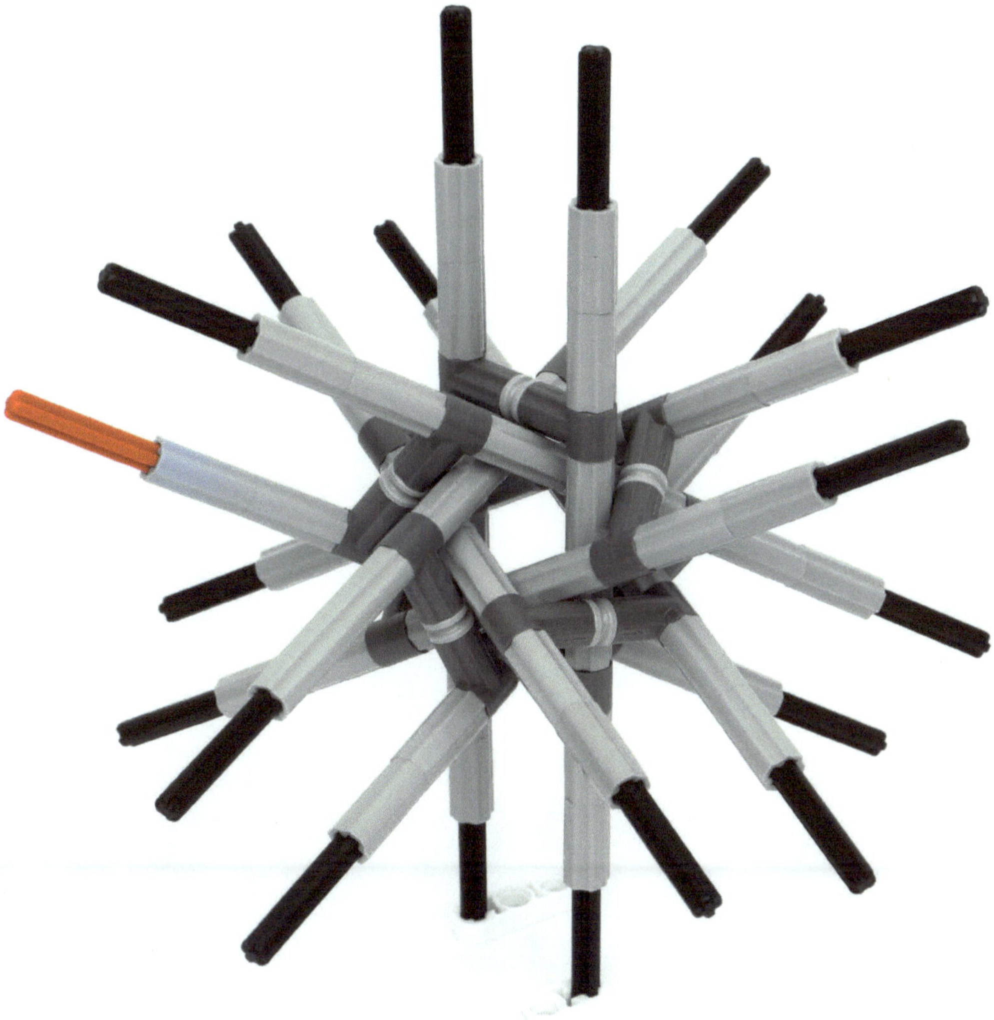

Intersection of 5 rectangles with extended edges.

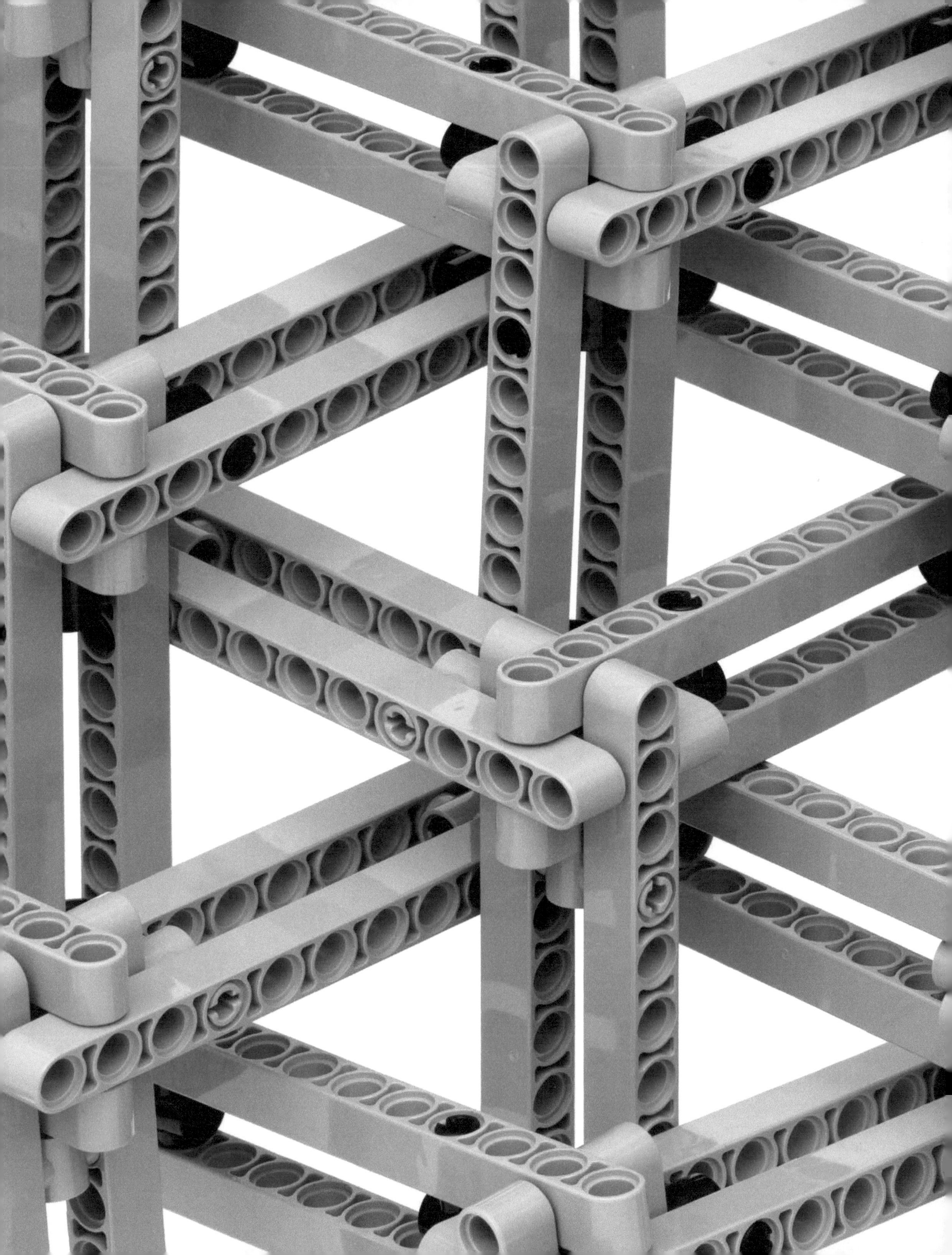

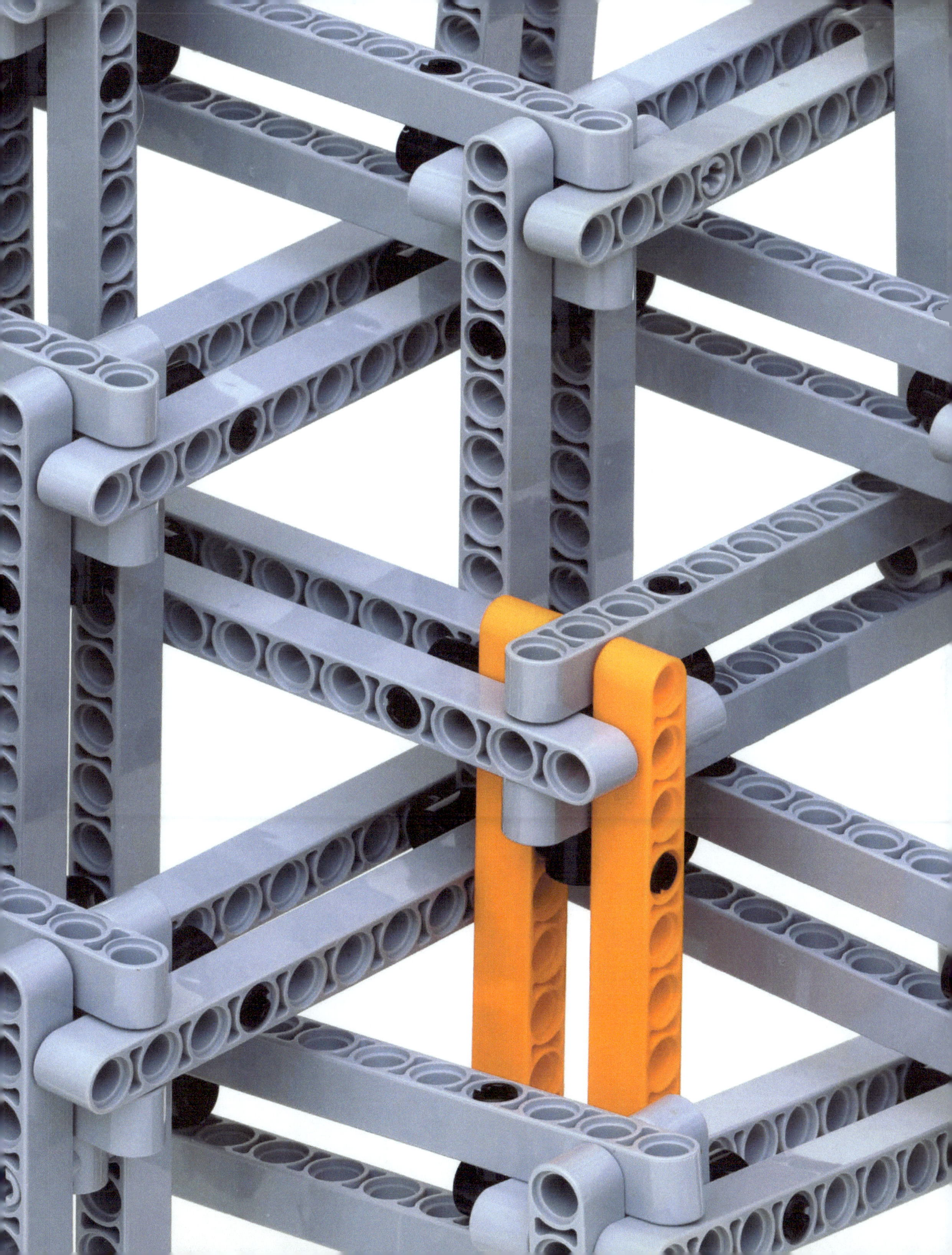

This 6 Helixes model is composed of 3 identical structures assembled at 120 degrees to each other.

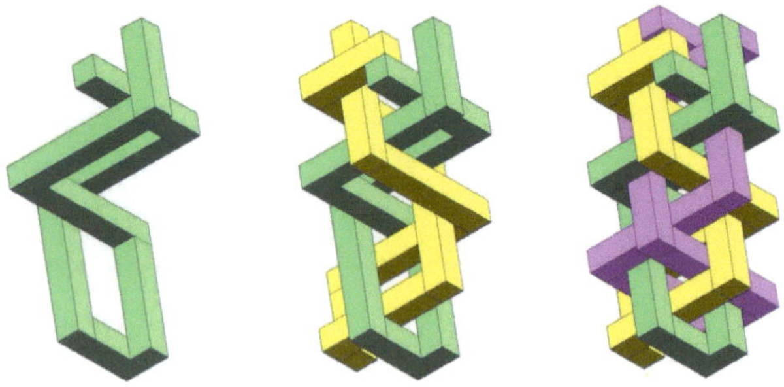

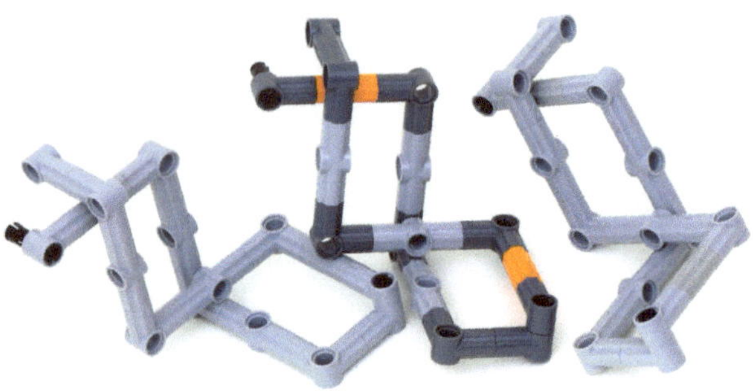

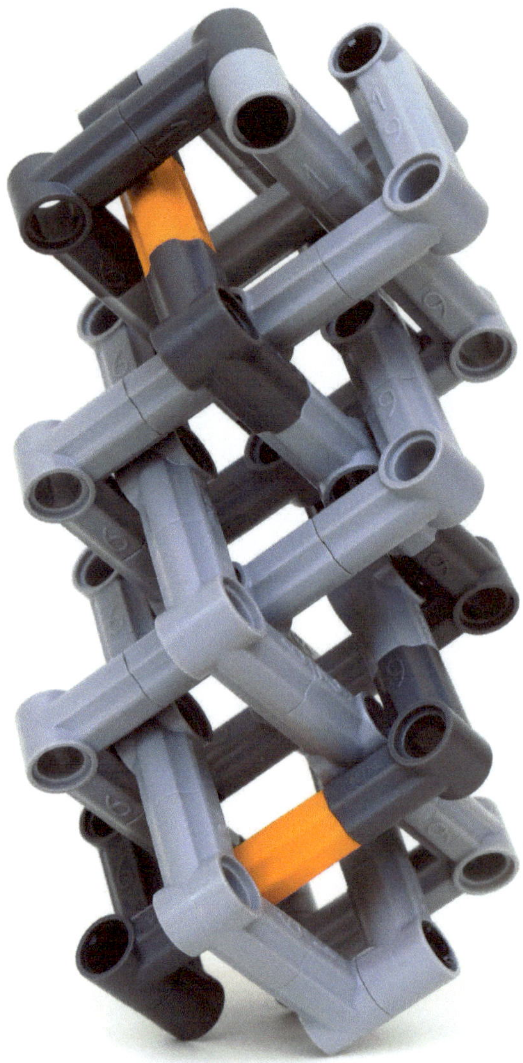

This is the same construction using 90 degree elbows and the lower picture is a half height version.

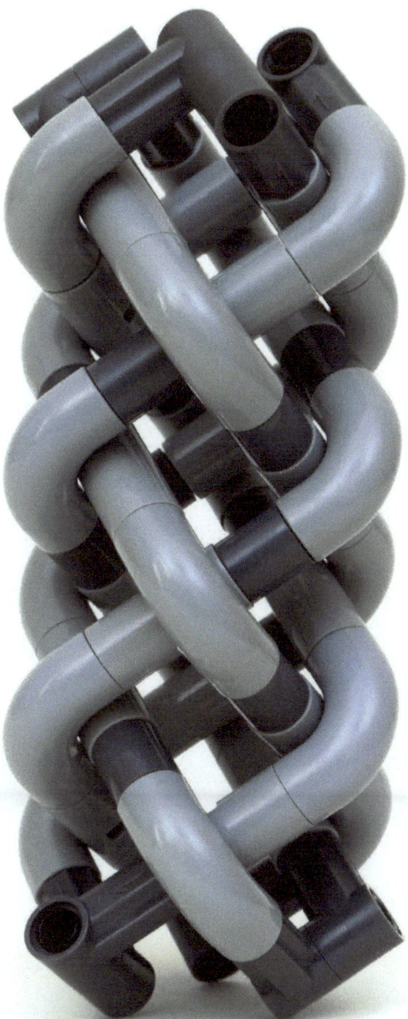

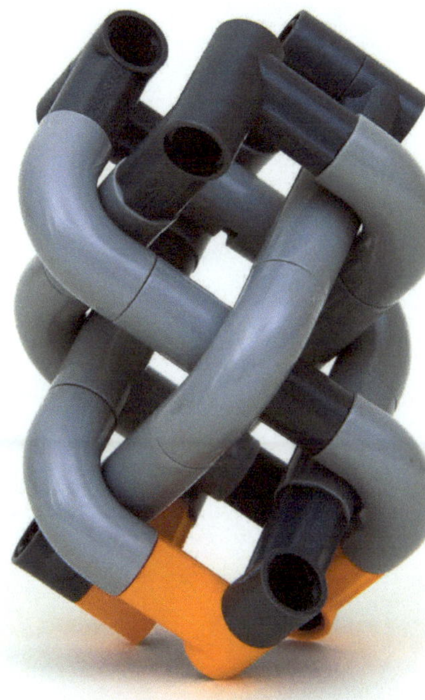

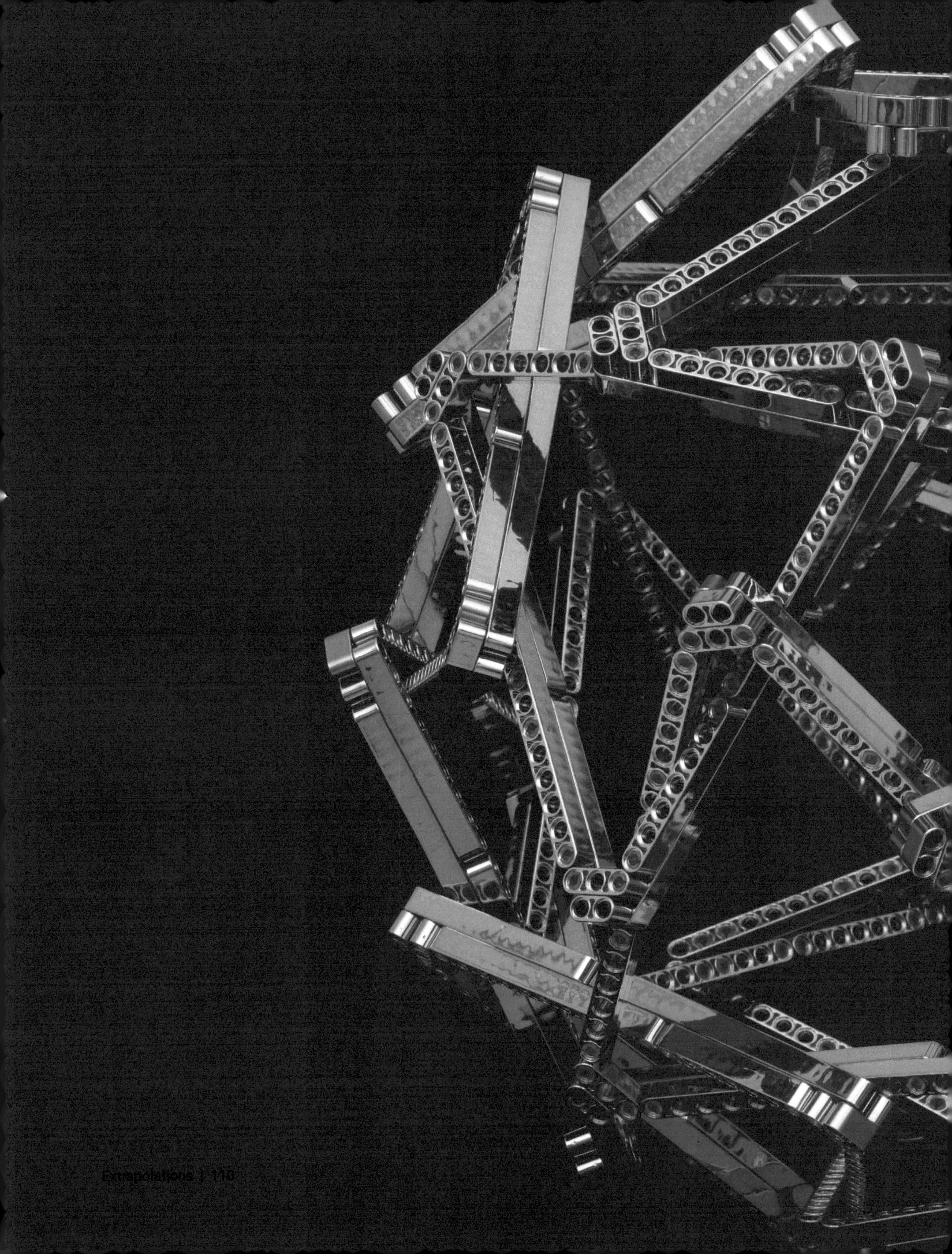

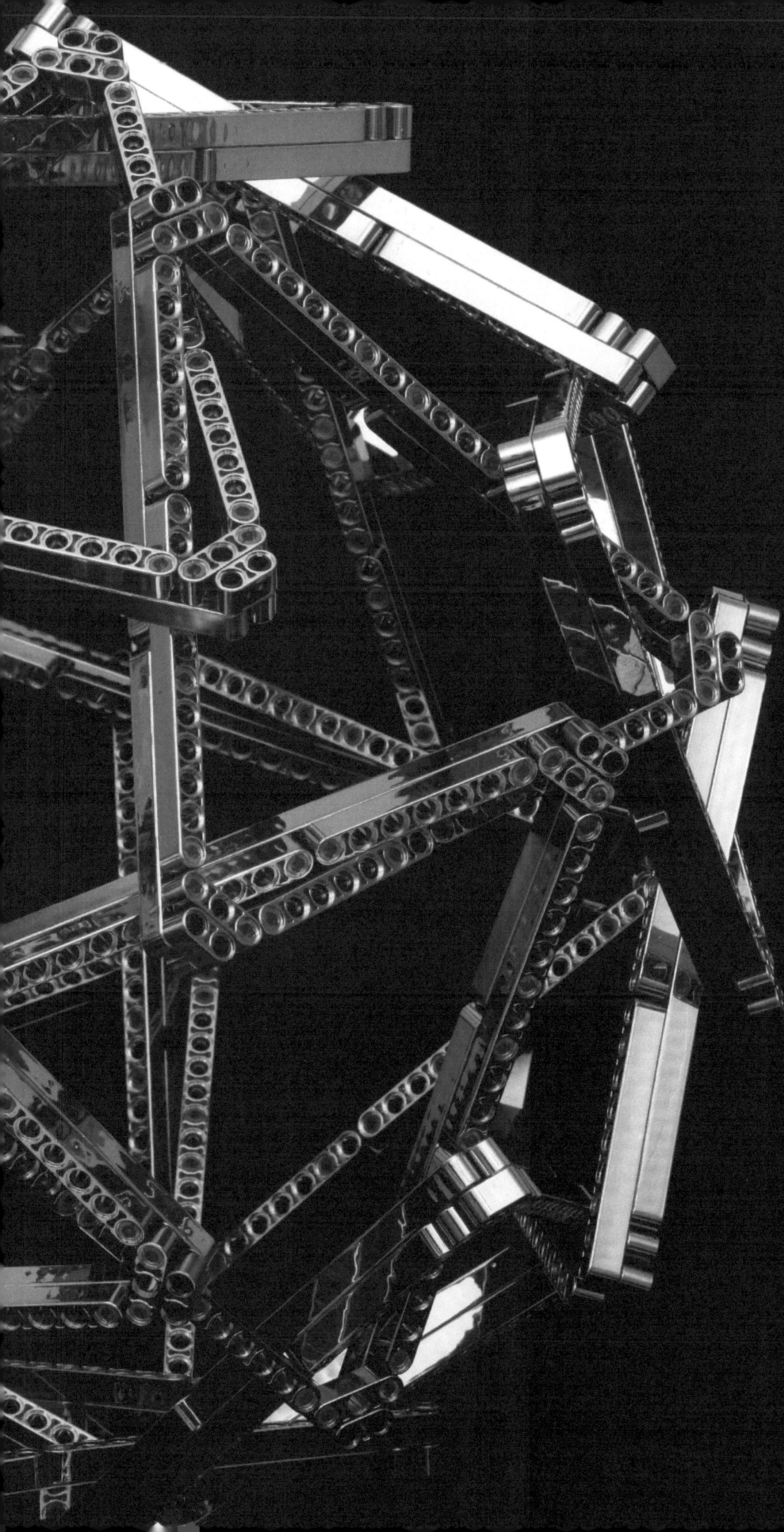

PARTS DESCRIPTION

According To Bricklink.com

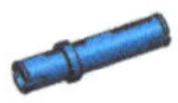

Technic, Pin 3L with Friction Ridges Lengthwise
Item No: 6558

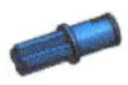

Technic, Axle Pin with Friction Ridges
Item No: 43093

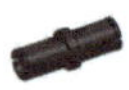

Technic, Pin with Friction Ridges. LengthwiseWITH Center Slots
Item No: 2780

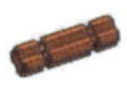

Technic, Axle 2 Notched
Item No: 32062

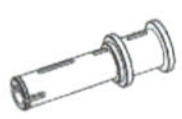

Technic, Pin 3L with Friction Ridges Lengthwise and Stop Bush
Item No: 32054

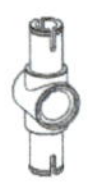

Technic, Pin 3L with Friction Ridges Lengthwise and Center Pin Hole
Item No: 87082

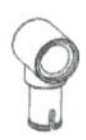

Technic, Pin with Friction Ridges Lengthwise and Pin Hole
Item No: 15100

Technic, Bush 1/2 Smooth
Item No: 4265c

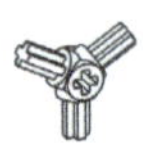

Technic, Axle Connector Hub with 3 Axles
Item No: 57585

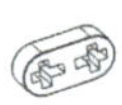

Technic, Liftarm 1 x 2 Thin
Item No: 41677

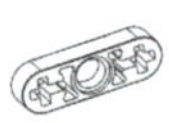

Technic, Liftarm 1 x 3 Thin
Item No: 6632

According To Bricklink.com

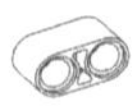

Technic, Liftarm 1 x 2 Thick
Item No: 43857

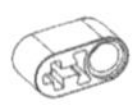

Technic, Liftarm 1 x 2 Thick with Pin Hole and Axle Hole
Item No: 60483

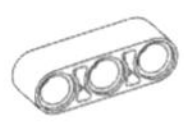

Technic, Liftarm 1 x 3 Thick
Item No: 32523

5

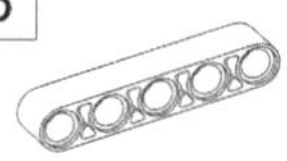

Technic, Liftarm 1 x 5 Thick
Item No: 32316

7

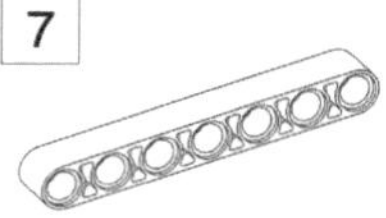

Technic, Liftarm 1 x 7 Thick
Item No: 32524

9

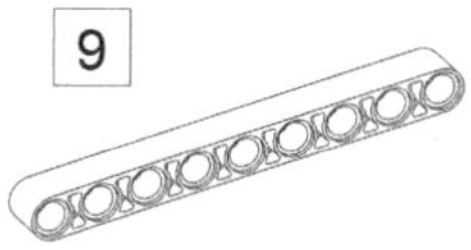

Technic, Liftarm 1 x 9 Thick
Item No: 40490

11

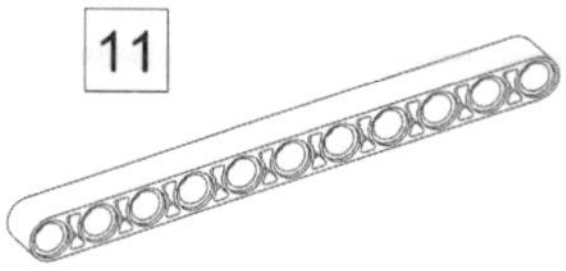

Technic, Liftarm 1 x 11 Thick
Item No: 32525

13

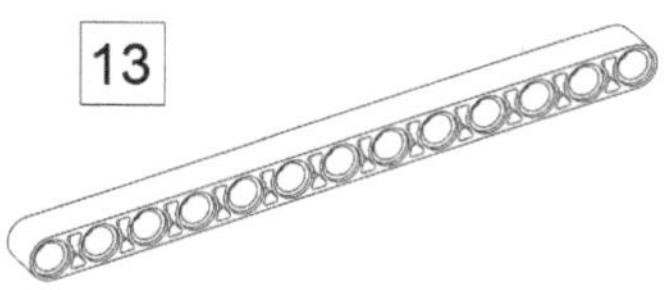

Technic, Liftarm 1 x 13 Thick
Item No: 41239

15

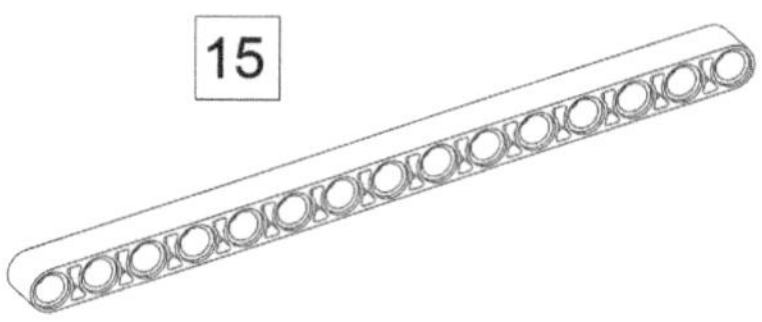

Technic, Liftarm 1 x 15 Thick
Item No: 32278

PARTS DESCRIPTION

According To Bricklink.com

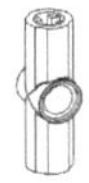

Technic, Axle and Pin Connector Angled #2 - 180 degrees
Item No: 32034

3

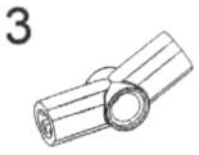

Technic, Axle and Pin Connector Angled #3 - 157.5 degrees
Item No: 32016

4

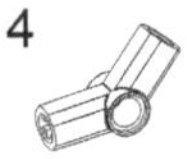

Technic, Axle and Pin Connector Angled #4 - 135 degrees
Item No: 32192

5

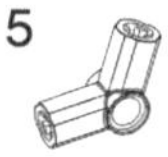

Technic, Axle and Pin Connector Angled #5 - 112.5 degrees
Item No: 32015

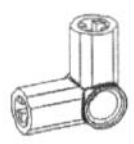

Technic, Axle and Pin Connector Angled #6 - 90 degrees
Item No: 32014

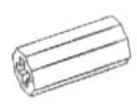

Axle Connector 2L (Smooth with x Hole + Orientation)
Item No: 6538c

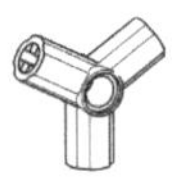

Technic, Axle and Pin Connector Triple
Item No: 10288

3

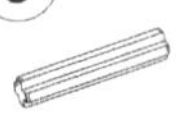

Technic, Axle 3
Item No: 4519

4

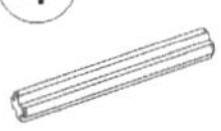

Technic, Axle 4
Item No: 3705

5

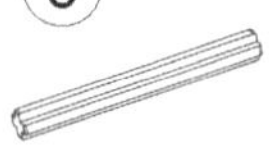

Technic, Axle 5
Item No: 32073

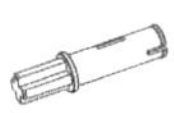

Technic, Axle Pin 3L with Friction Ridges Lengthwise and 1L Axle
Item No: 11214

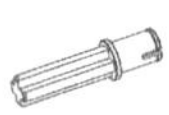

Technic, Axle Pin 3L with Friction Ridges Lengthwise and 2L Axle
Item No: 18651

According To Bricklink.com

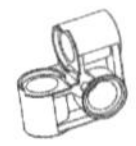

Pin Connector Perpendicular 2 x 2 Bent
Item No: 44809

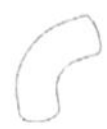

Brick, Round 1 x 1 d. 90 Degree Elbow - No Stud - Type 2 - Axle Hole
Item No: 25214

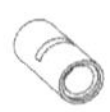

Technic, Pin Connector Round 2L with Slot
Item No: 62462

Technic, Pin Connector Round 2/3 L
Item No: 18654

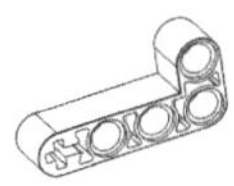

Technic, Liftarm 2 x 4 L-Shape Thick
Item No: 32140

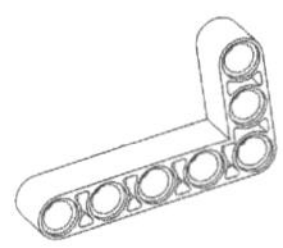

Technic, Liftarm 3 x 5 L-Shape Thick
Item No: 32526

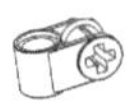

Technic, Axle and Pin Connector Perpendicular
Item No: 6536

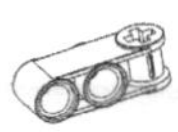

Technic, Axle and Pin Connector Perpendicular 3L with 2 Pin Holes
Item No: 42003

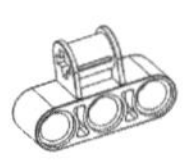

Technic, Axle and Pin Connector Perpendicular Triple
Item No: 63869

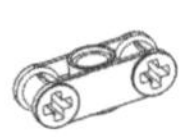

Technic, Axle and Pin Connector Perpendicular 3L with Center Pin Hole
Item No: 32184

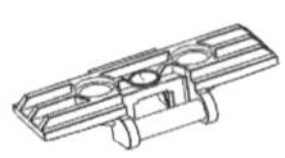

Technic, Link Tread Wide with Two Pin Holes
Item No: 57518

GLOSSARY

Borromean link configuration - A configuration similar to Borromean ring - In mathematics, the Borromean rings consist of three topological circles which are linked but where removing any one ring leaves the other two unconnected.
In other words, no two of the three rings are linked with each other as a Hopf link, but nonetheless all three are linked. The Borromean rings are one of a class of such links called Brunnian links.
Borromean rings (31 January 2020). In Wikipedia.Retrieved from https://en.wikipedia.org/wiki/Borromean_rings

Brunnian link - In knot theory, a branch of topology, a Brunnian link is a nontrivial link that becomes a set of trivial unlinked circles if any one component is removed. In other words, cutting any loop frees all the other loops (so that no two loops can be directly linked).
Brunnian link (29 December 2019). In Wikipedia. Retrieved from https://en.wikipedia.org/wiki/Brunnian_link

CAD - Computer Aided Design

Extrapolation - further design variations of the existing model.

Hexapod type - a model having 6 corners or vertices.

Interlocking - an interlocking sculpture having their adjacent parts mutually locking each other in place.

Inverse version - a variation of the 'normal' version. Many of the geometric sculptures that are derived from the polyhedra may have 2 or more intersecting configurations. The common intersecting configuration is the hollow faces translate and rotate simultaneously towards the center. The other solution is translating and rotating towards the outermost edges of the polyhedra. The outer version I use
the term inverse version.

Orthogonally - to position adjacent parts or structures perpendicular to one another.

Topologically - from the word topology.
In mathematics, topology (from the Greek τόπος, 'place', and λόγος, 'study') is concerned with the properties of a geometric object that are preserved under continuous deformations, such as stretching, twisting, crumpling and bending, but not tearing or gluing.
Topology. (8 February 2020). In Wikipedia. Retrieved from https://en.wikipedia.org/wiki/Topology

Tripod type - a model having 3 corners or vertices.

Valknut - The valknut (coined from Old Norse valr, "slain warriors" and knut, "knot") is a symbol consisting of three interlocked triangles. It appears on a variety of objects from the archaeological record of the ancient Germanic peoples.
Valknut. (29 December 2019). In Wikipedia. Retrieved from https://en.wikipedia.org/wiki/Valknut

CPSIA information can be obtained
at www.ICGtesting.com
Printed in the USA
LVHW070427060721
691874LV00001B/19

* 9 7 8 1 5 4 3 7 6 4 8 3 3 *